Forgotten Americans
who served
in the War of 1812

Also by Eric Eugene Johnson

American Prisoners of War Held at Quebec During the War of 1812:
8 June 1813 – 11 December 1814

Ohio and the War of 1812:
A Collection of Lists, Musters and Essays

Ohio's Regulars in the War of 1812

Ohio's Black Soldiers Who Served in the Civil War

American Prisoners of War held in Montreal and Quebec
During the War of 1812

American Prisoners of War Paroled at Dartmouth, Halifax,
Jamaica and Odiham during the War of 1812

Black Regulars in the War of 1812

American Prisoners of War Held at Dartmoor
during the War of 1812

# Forgotten Americans
# Who Served
# in the
# War of 1812

Eric Eugene Johnson

Society of the War of 1812
in the
State of Ohio

HERITAGE BOOKS
2018

# HERITAGE BOOKS
## *AN IMPRINT OF HERITAGE BOOKS, INC.*

**Books, CDs, and more—Worldwide**

For our listing of thousands of titles see our website
at
www.HeritageBooks.com

Published 2018 by
HERITAGE BOOKS, INC.
Publishing Division
5810 Ruatan Street
Berwyn Heights, Md. 20740

International Standard Book Number
Paperbound: 978-0-7884-5826-2

# Acknowledgments

I wish to thank Dr. Rose Mary S. Rudy, Ph.D. and E. Paul Morehouse for helping me make this book a reality. Without the help and support of friends, life would be difficult!

# Contents

# Introduction

Many books have been written on the War of 1812 but few deal with the sacrifices of the common people and on the logistics of this conflict. Book after book deals with the battles, and of the men who led the armies and who made the decisions from the safety of Quebec City and Washington, D.C.

There are hundreds of stories of men, women and children who were directly affected by this war on a day-to-day basis. Women served as washerwomen, matrons and servants to the men in the field. American children died in prisoner ships off Quebec City. A Tennessee father enlisted in the army with three of his sons, all under age, and all served as soldiers. Three African American brothers enlisted together in the army. They hailed from North Carolina as free men who wanted to fight for their country. And the list goes on!

The logistics of the war can be as interesting as the battles. How were the regiments organized and raised? What was the Corps of Artificers? Who were the Sea Fencibles? Why did we surrender Fort Sullivan in the District of Maine? These and many more stories are going to be revealed in this book.

Introduction

# The men, women and children

The men, women and children

## American women who served in the War of 1812

We tend to overlook or may not be aware of the role that women played during the War of 1812 in support of our military. Women worked in the "cottage" industries sewing uniforms and rolling cartridges for our boys in blue. They also made buckskin jackets and prepared rations for the militia. A hand-full of women actually served with our troops on the battle lines during this war.

These women were not camp followers, but members of the regular army and the militia forces who served as washerwomen, servants, matrons and nurses. Many women, along with their children, became prisoners of war. One woman died in the British prison camp in Montreal, Canada, while six American children died at Quebec, Canada, on board the prison ships. Another woman, who was disguised as a man, was sent to the prison camp at Quebec after being captured, and she was later released. Other women were captured on the high seas on American merchant ships by the Royal Navy and interned in prison camps in England.

Under the Congressional Act of 16 March 1802 up to four women could be hired to serve as washerwomen in each of the army's companies. This law also permitted women to serve as matrons and nurses in military hospitals, and as servants for commissioned army officers.

The Congressional Act of 6 July 1812 placed the women on equal footing with privates, that is, the women would receive the same pay, and the same basic allowances for clothing and subsistence as a common soldier. They were also permitted to receive one ration per day. Most of these women were the wives of soldiers. They were invaluable in the day-to-day operations of a company or a hospital.

Eliza Romley is the only American women known to have been captured during the War of 1812 as a combatant. Disguised as a seaman, Eliza served onboard the U.S. Sloop *Growler* on Lake Champlain. She was captured on 3 June 1813 when this ship fell into British hands. She was released twenty-two days later from the prison camp at Quebec after her true identity was discovered.

Romley is not a common name found during this time period in New England nor in New York. There is a possibility that she is a member of the Bromley family of Danby, Vermont. Her surname may have been misspelled by the British clerk in Quebec. Eliza Romley may also be an alias and therefore the true identity on this woman may never be known.

A clothing report for the 24th Regiment of U.S. Infantry for American prisoners of war who were held in Montreal lists the death of an American women. The report states that Mrs. Bledsoe, the wife of a private in the 24th Infantry died on 22 April 1814. Mrs. Bledsoe was probably a washerwomen in this regiment, who was captured when Fort Niagara, NY, fell to the British on 19 December 1813. The 24th Infantry was raised in Tennessee.

The identify of this women was revealed two days later in a letter written on 24 April 1814 by Robert Gardner, the agent for the American prisoners of war in Canada, to his superior, John Mason, the Commissary General of Prisoners. Gardner states, "To the men I have paid each a small sum, $4; to one man of the 24 regiment whose wife died here two days since, $9 (Joseph Bledsoe)". Besides receiving new clothing, the American prisoners of war also received $4 each from Gardner, however, Bledsoe received an additional $5 because of the death of his wife.

After being released by the British at war's end, Bledsoe relocated to Allegany County, New York, where he remarried and raised a family. *The History of Allegany County, New York*, states "A man who will long be remembered in the town was Joseph Bledsoe (a son-in-law of John Teater), a Virginian by birth, who after serving in the war of 1812 removed to Amond at its close, having been a prisoner of war at Montreal at the time peace was declared. In 1820 he came to Independence, and settled on lot number 81, north of the location of John Teater, whose daughter he had married in 1816."

Bledsoe applied for a pension after the war, which was rejected. He states in his pension papers that he was wounded during the Battle of Fort Niagara and that he was made a prisoner of war by the British. He makes no mention of his first wife in his application, even though she had died in captivity in Montreal.

During the two sieges of Fort Meigs, Ohio, between 28 April and 9 May 1813, and again between 21-28 July 1813, there were six infantry, one artillery and one light dragoon companies of the U.S. Army stationed at this fort. There could have been up to 32 women within this fort serving as washerwomen at the time of the siege. During this siege, the women were probably assisting in the fort's hospital, or they were carrying food and water to the men who were manning the walls, blockhouses and artillery ramparts.

A general order, issued at Fort Meigs on 1 August 1813 states, "Any married woman who has or shall abandon her husband and be found strolling about camp or lodging in the tents of other men shall be drum'd out of camp." Although women may not have been present during the two sieges of Fort Meigs, there were women serving at this fort during the summer of 1813.

Most company musters and rosters don't list the names of these women who were hired to do the laundry of the men. Three recruiting rosters, however, do list the names of some of these women who served in Ohio. On a roster of recruits being sent to Chillicothe from Zanesville on 9 August 1812, two women are listed as washerwomen: Polly Waters and Betsey Laurence. Chillicothe was the recruiting rendezvous for the 19th Regiment of U.S. Infantry, where recruits were organized into companies and then sent to the front to join the rest of the regiment.

On 18 September 1812, another detachment of recruits was sent to Chillicothe from Zanesville. On this roster two more women are listed: Susannah Bright, the wife of George Bright, and Fanny Stultz, the wife of Adam Stultz. Adam would be killed on 5 May 1813 during the first siege of Fort Meigs. Fanny may have been with her husband at Fort Meigs and she may have witnessed his death. The Stultz's were from Muskingum County, Ohio.

Captain Daniel Cushing of Ohio commanded the artillery company from the 2nd Regiment of Artillery, which was stationed at Fort Meigs. Mrs. Aaron Haning is listed on a recruiting detachment report for this company that rendezvoused at Chillicothe on 29 October 1812.

## Children of American POWs who died at Quebec during the War of 1812

God bless the six American children, all three years and under, who died on the British prison ships off the City of Quebec during the War of 1812. Women and children were not always protected from the horrors of war and this war was no exception. There were American women and children serving as prisoners of war along with their fathers and husbands from 1812 through 1815 in British prisoner of war facilities. They suffered the same fate and hardships as their men. Our history has forgotten them!

During the war the British captured three major American forts: Fort Mackinac, Territory of Michigan, on 17 July 1812, Fort Detroit, Territory of Michigan, on 16 August 1812 and Fort Niagara, New York, on 19 December 1813. All of the occupants in these forts became prisoners of war: the officers, the soldiers, their women and children. These women, who lost their children, were the wives of soldiers. Most were probably working as washerwomen in their husband's companies.

An article in the weekly newspaper, *Quebec Weekly Chronicle*, dated 18 July 1889, by N. LeVasseur lists 90 Americans who died on the prison ships off Quebec during the war. LeVasseur, first name not listed, extracted the list from the registers of the English Metropolitan Church of Quebec. It appears that all of the Americans were given Christian burials in this church's graveyard.

A second source confirming the identity of the six children is the diary of Surgeon's Mate James Reynolds of the 4th Regiment of U.S. Infantry. His diary, *Journal of an American Prisoner at Fort Malden and Quebec in the War of 1812*, was published in Quebec in 1909.

Reynolds worked as an army doctor caring for the men and their families on the prison ships. He was interned on prison ship number 160, which is probably the British transport *Malabar*. In his diary, he lists five ships by their numbers in which the Americans were interned. On his ship were twelve women.

Both sources contain numerous spelling errors and the surnames were spelled phonetically. Using the U.S. Army's *Register of Enlistments, 1798-1914* and the *Records Relating to War of 1812 Prisoners of War* from the National Archives in Washington, DC (Microfilm Publication M2019 in Record Group 94), the corrected list of children can be listed.

1) John Stoner, son of Sergeant John and Mary Stoner, 1st Regiment of U.S. Infantry, age 2 months, died on 6 October 1812.

2) Parmelia Perry, daughter of Corporal Calvin and Anna Perry, 1$^{st}$ U.S. Infantry, age 2 years, died on 8 October 1812.
3) John Guiles, son of Private Joseph and Eleanor Guiles, 1$^{st}$ U.S. Infantry, age 3 years, died on 12 October 1812
4) Jane Whitelock, daughter of Sergeant John Whitelock, 1$^{st}$ U.S. Infantry, age 2 years, died on 12 October 1812.
5) Lucinda Weir, daughter of Private David and Ann Weir, 4$^{th}$ Regiment of U.S. Infantry, age 13 months, died on 14 Oct 1812.
6) Stephen Ingalls, son of Private Amos and Abigail Ingalls, 4$^{th}$ U.S. Infantry, age 2 years and 6 months, died on 14 Oct 1812.

These Americans were captured after the fall of Fort Detroit and transported to Quebec. They were later released on parole and sent to Boston, arriving on a number of ships between 24 November and 30 November 1812. The men were then stationed at Fort Independence, in Boston harbor, until they were exchanged for British prisoners of war. They were then sent back to their regiments.

The British captured two artillery companies, a company from the 1st Regiment of U.S. Infantry plus a detachment of recruits, and the 4th Regiment of U.S. Infantry at Fort Detroit. The militiamen from Michigan and Ohio, who were also captured, were paroled and sent home.

Ending on a happier note! Mrs. Andrew gave birth to a son on 4 October 1812 while on a prison ship. The husband is unknown but it is either E. T. Andrews, Otis Andrews or William Andrews, all from the 4th U.S. Infantry. Mother and child appear to have survived since they are not noted on LeVasseur's list.

## Definitions

3$^{rd}$ U.S. Military District – Lower New York and eastern New Jersey

4$^{th}$ U.S. Military District – Western New Jersey, Pennsylvania and Delaware

Matron – A washerwomen in a hospital who laundered the sheets and made the beds.

MV – Merchant vessel

Nurse – A person trained to care for the wounded and the sick in army hospitals. This position was normally reserved for men during the War of 1812.

Servant – A civilian who was hired as a waiter. These servants could be a personal friend, a slave or a women.

Waiter – A private in the army who was assigned as a servant to a commissioned officer.

Washerwomen – A women who laundered the clothes of soldiers.

## The Women and Children

-----, Lucy - Camp follower - 7th U.S. Infantry - Present at Tchefuncte, LA.

-----, Tokey - Indian Squaw - Woman - Present at New Orleans, LA, Apr 1814.

Adams, Sarah - Washerwoman - 42nd U.S. Infantry - Company: Captain Jonathan Roberson - Inspection return, Philadelphia, PA, 28 Feb 1815.

Anderson, Sally - Washerwoman - 19th Regiment, Virginia Militia (Lieutenant Colonel John Ambler).

Andrews, Mrs. - Civilian - 4th U.S. Infantry - Prisoner of war at Quebec, Canada; gave birth to a son on 4 Oct 1812; wife of E. T., Otis or William Andrews.

Artis, Mary - Military hospital patient - Admitted to hospital, 7 Sep 1814 and discharged from hospital, 30 Sep 1814.

Baker, Judith - Servant - 7th Regiment, Virginia Militia (Lieutenant Colonel David Saunders).

Baker, Mary - Washerwoman - 40th U.S. Infantry - Company: Captain Mathew Sanborn - Inspection return, Fort Constitution, MA, 28 Feb and 21 Apr 1815, present.

Beaty, Elizabeth - Nurse - 3rd U.S. Military District - Monthly return, U.S. General Hospital Bella Vero, 21 January 1815, discharged 26 January.

Bentley, Susan – U.S. Ordnance Department - Company: First Lieutenant Nehemiah Baden's Detachment - Inspection return, 23 Feb and 30 Apr 1815, Greenleafs Point, Washington, DC, present.

Betsy Armstrong - Nurse - Hospital Attendants, Pennsylvania.

Bledsoe, Mrs. - Civilian - 24th U.S. Infantry - Died at Montreal, Canada, on 22 Apr 1814 while a prisoner of war; wife of Private Joseph Bledsoe.

Booth, Ann - Nurse - Nurse in General Hospital, 3rd U.S. Military District.

Bowers, Nancy - 32nd U.S. Infantry - Company: Captain William Smith - Monthly return, 31 Oct 1813.

Briggs, Mary - Washerwoman - 1st Regiment, Massachusetts Militia (Lieutenant Colonel Joseph Valentine).

Briggs, Susannah - Washerwoman - 1st Regiment, Massachusetts Militia (Lieutenant Colonel Joseph Valentine).

Bright, Susannah - Washerwoman - 19th U.S. Infantry - Company: Captain Wilson Elliott - Wife of George Bright.

Brower, Betsey - Private - 1st U.S. Volunteer Regiment - Company: Captain Burrows - Hospital return 1812 to 1815, received at U.S. General Military Hospital, City of New York, 26 April 1813, disease pneumonia, present, 30 April 1813; discharged 3 May 1813, cured.

Carter, Mary - Washerwoman - 40th U.S. Infantry - Company: Captain Robert Neal, Junior - Inspection return, Fort McClary, ME, 28 Feb and 30 Apr 1815, present.

Cavender, Sarah - Servant – U.S. Corps of Engineers - Enlistment date: 1 Dec 1814 - Monthly return, U.S. Military Academy, West Point, NY; private servant to Captain Samuel Perkins, Quartermaster, Feb-Apr 1815.

Chamberlain, Hannah - Nurse in U.S. General Hospital, 3rd U.S. Military District.

Clark, Miss - Washerwoman – U.S. Corps of Artillery - Company: Captain Ichabod Crane - Present at Sackets Harbor, NY, 28 Feb 181.

Crump, ----- - Washerwoman - 41st U.S. Infantry - Company: Captain Francis Allyn - Inspection Return at Hurl Gate, N.Y., 1 Mar 1815.

Drew, ----- - Waiter - 2nd Regiment, Massachusetts Militia (Colonel Salem Town, Jr.).

Fell, Tamey - Hospital Nurse - Major General Nathaniel Watson's Division, Pennsylvania Militia.

Gray, Amey - Servant - 5th Regiment, Virginia Militia.

Greenlee, Mrs. - Washerwoman - 45th U.S. Infantry - Company: First Lieutenant Henry Snows' Detachment - Inspection return, Phippsburg, ME, 28 Feb 1815.

Greeny, Mrs. - Washerwoman - 8th Regiment, New York Militia (Lieutenant Colonel Thomas Miller).

Griffin, Mrs. - Washerwoman – U.S. Sea Fencibles - Company: Captain Jonathan Du Bose – Recruiting

Report, Willow Grove, SC, 11 Jun 1814.

Guiles, Eleanor - Civilian - 1st U.S. Infantry - Prisoner of War at Quebec, Canada; wife of Private Joseph Guiles.

Guiles, John - Child - 1st U.S. Infantry - Age: 3 years - Died at Quebec, Canada, on 12 Oct 1812 while a prisoner of war; parents: Private Joseph and Eleanor Guiles.

Hammond, Cynthia - Washerwoman – U.S. Corps of Artillery - Company: Captain John Walbach - Inspection return, Fort Constitution, MA, 30 Apr, 30 Jun, 31 Aug, 31 Oct and 31 Dec, 1815, present.

Haning, Mrs. - Washerwoman - 2nd U.S. Artillery - Company: Captain Daniel Cushing - Wife of Aaron Hanning.

Harker, Mary - 14th U.S. Infantry - Admitted to hospital, 7 Sep 1814 and discharged from hospital, 30 Sep 1814.

Harrison, Elizabeth - Private - 10th U.S. Infantry - Company: Captain Thomas Nelson - Admitted to hospital, 7 Sep 1814 and discharged from hospital, 30 Sep 1814.

Harriss, Ann – U.S. Corps of Engineers - Enlistment date: 1 Jun 1814 - Monthly return, U.S. Military Academy, West Point, NY, 20 Sep 1814, private servant to Adam Empie, Chaplain and Professor of Ethics; present 28 Feb 1815 and 30 Apr 1815.

Hazelton, Ann - Nurse - 3rd U.S. Military District - Monthly returns, U.S. General Hospital Bellevue, 31 Jul 1815.

Henry, Catharine - Matron - Fort Niagara, NY - Company: Hospital - Medical staff at Niagara.

Henry, Polly - Matron - Fort Niagara, NY - Company: Hospital - Medical staff at Niagara.

Hetick, Mrs. - Washerwoman - 41st U.S. Infantry - Company: Captain Francis Allyn - Inspection return, Hurl Gate, NY, 1 Mar 1815.

Hickman, Rosanna - Matron - Fort Niagara, NY - Company: Hospital - Monthly return, medical stall, Niagara, Dec 1814, present.

Hollowell, Betsey - Nurse - Hospital Attendants, Pennsylvania.

Holmes, Elizabeth - 10th U.S. Infantry - Company: Captain George Vashon - Admitted to hospital, 7 Sep 1814 and discharged from hospital, 30 Sep 1814.

Hood, Margaret - Private - 5th U.S. Infantry - Company: Captain John Gassaway - Admitted to hospital, 7 Sep 1814 and discharged from hospital, 30 Sep 1814.

Humphrey, Mrs. - 7th U.S. Infantry - Order book, 4 Jan 1814 to 5 Feb 1815, present.

Hunt, Peggy I. - Servant - 1st Regiment, Virginia Militia (Colonel James Byrne).

Ingalls, Abigail - Civilian - 4th U.S. Infantry - Prisoner of war at Quebec, Canada; wife of Private Amos Ingalls.

Ingalls, Stephen - Child - 4th U.S. Infantry - Age: 2 years and 6 months - Died at Quebec, Canada, on 14 Oct 1812 while a prisoner of war; parents: Private Amos and Abigail Ingalls.

Johnson, Eve - Hospital matron - 2nd U.S. Artillery - Company: Captain Alexander J. Williams - Enlistment date: 7 Jul 1813 - Monthly return of the 4th Military District and the monthly return of Fort Mifflin, PA, 30 Oct 1813, present.

Johnson, Hetty - Nurse - Hospital Attendants, Vermont.

Jones, Becky - Nurse - Hospital Attendants, Pennsylvania.

Jones, Letticia - Waiter - 16th U.S. Infantry - Company: Captain R. Patterson - Monthly return, 1813 and 1814, private waiter for Lieutenant Bunting; absent on duty at Philadelphia.

Jones, Polly - 2nd U.S. Volunteer Regiment - Company: Captain Samuel Perlee - Born: NY - Hospital return, 1812 to 1815, received at U.S. Military Hospital, Greenwich, NY, 3 Aug 1813; child died of cholera infantum on 5 Aug 1813; Polly Jones discharged 6 Aug 1813.

Kanady, Mary - Washerwoman - 40th U.S. Infantry - Company: Captain Robert Neal Junior - Inspection return, Fort McClary, ME, 28 Feb and 30 Apr 1815, present.

Kelly (O'Kelly), Elizabeth - Nurse - Hospital Attendants, Pennsylvania.

Kelly, Betsey - Nurse - Hospital Attendants, Pennsylvania.

Kelly, Margaret - Rifle Cantonment, NY - Orders dated Rifle Cantonment, 9 Feb 1815; tried by Detachment Court-martial; charge - misconduct; sentenced to be dismissed from camp but on recommendation of court, and in consideration of her situation, sentence remitted.

Kinnaer, Darias - Washerwoman - 40th U.S. Infantry - Company: Captain Robert Neal Junior - Stationed at Fort McClary, ME.

Laurence, Betsey - Washerwoman - 19th U.S. Infantry.

Leonard, Catherine - Matron - Fort Niagara, NY - Company: Hospital - Medical staff at Niagara.

Loper, Josie - Nurse - 61st Regiment, Virginia Militia (Lieutenant Colonel Leavin Gayle).

Lydia McCully - Matron - Hospital Attendants, Pennsylvania.

Manley, Milley - Nurse - 61st Regiment, Virginia Militia (Lieutenant Colonel Leavin Gayle).

Marcey, Nancy - Servant - 4th Regiment, Virginia Militia.

Marvin, Deborah - Washerwoman - 40th U.S. Infantry - Company: Captain Robert Neal Junior - Inspection return, Fort McClary, ME, 28 Feb 1815 and 20 Apr 1815, present.

McCammon, Elizabeth - Nurse in U.S. General Hospital, 3rd U.S. Military District from 1 Nov 1813 to 1 Feb 1814 and 1 Aug 1814.

McCannan, Mary - Nurse - Hospital Attendants, Pennsylvania.

Morgan, Sally - Washerwoman – U.S. Corps of Artillery - Company: Captain John Walbach - Inspection return, 28 Feb, 30 Apr, 30 Jun, 31 Aug and 31 Dec 1815, present.

Morris, Mary - Hospital Nurse - 1st Regiment, Riflemen, Pennsylvania Militia (Colonel Thomas Humphrey).

Morris, Mary - Nurse - Hospital Attendants, Pennsylvania.

Moss, Abigail - Servant – U.S. Corps of Engineers - Enlistment date: 2 Jul 1814 - Monthly return, U.S. Military Academy, West Point, NY; private servant to Captain Samuel Perkins, Quartermaster, Sep 1814.

Murphy, Sarah - Nurse - Hospital Attendants, Pennsylvania.

Nickerson, Betsey - Waiter - 1st Regiment, Massachusetts Militia (Colonel Joseph Dudley).

Norton, Margaret (or Mary) - Women - 32nd U.S. Infantry - Company: Captain William Smith - Monthly return, 31 Oct 1813.

Ormsby, Nancy - Washerwoman – U.S. Corps of Artillery - Company: Captain John Walbach - Fort

Constitution, MA; Feb 1814-Dec 1815.

Payett, Mary - Washerwoman - New York Militia - Company: Captain Lemon Foot's Artillery Company.

Pea, Mrs. John - 5th U.S. Infantry - Company: Lieutenant Edward Rephan's Detachment - Inspection return, 28 Feb 1815.

Perry, Anna - Civilian - 1st U.S. Infantry - Prisoner of War at Quebec, Canada; wife of Corporal Calvin Perry.

Perry, Mrs. - 43rd U.S. Infantry - Company: Captain Henry Garrett - Inspection return, Fort Hampton, 1 Aug 1815.

Perry, Parmelia - Child - 1st U.S. Infantry - Age: 2 years - Died at Quebec, Canada, on 8 Oct 1812 while a prisoner of war; parents: Corporal Calvin and Anna Perry.

Phillippa, Cecilia - 32nd U.S. Infantry - Company: Captain William Smith - Monthly return, 31 October 1813.

Phillips, Martha - Matron - Fort Niagara, NY - Company: Hospital - Medical staff at Niagara, Dec 1814.

Pinkerton, Mrs. - 41st U.S. Infantry - Company: Captain Francis Allyn - Inspection return, Hurl Gate, New York, NY, 1 Mar 1815.

Pocy, John - Passenger - Privateer Volant - Age: 11 - Captured on 26 Mar 1813 and paroled at Dartmouth, England, on 5 Apr 1813 (Prisoner of War).

Pocy, Joseph - Passenger - Privateer Volant - Age: 13 - Captured on 26 Mar 1813 and paroled at Dartmouth, England, on 5 Apr 1813 (Prisoner of War).

Powell, Juliet - Washerwoman - Inspection return for Greenleafs Point, NY, 28 Feb 1815.

Punch, Ann - Hospital Nurse - Major General Nathaniel Watson's Division, Pennsylvania Militia.

Ray, Sarah - Washerwoman - 10th U.S. Infantry - Company: Captain George Vashon - Admitted to hospital, 7 Sep 1814 and discharged from hospital, 30 Sep 1814.

Richards, Elizabeth - Matron - Greenbush Cantonment, NY - Company: Hospital - Enlistment date: 17 May 1814 - Greenbush Cantonment, NY, Feb-Apr 1814.

Risbrough, Edward - Child - MV Mary Ann - Captured on 9 Aug 1812 and paroled at Oliham, England, on 3 Nov 1812 (Prisoner of War); sent to Chatham, England, on 8 Mar 1813; parents: John and Margaret Risbrough.

Risbrough, Margaret - Passenger - MV Mary Ann - Captured on 9 Aug 1812 and paroled at Oliham, England, on 3 Nov 1812 (Prisoner of War); sent to Chatham, England, on 8 Mar 1813; wife of John Risbrough (passenger).

Rogers, Eliz. - Washerwoman - 68th Regiment, Virginia Militia.

Rogers, Mrs. - Washerwoman - Woman - Inspection return, Fort Hampton, NC, 1 Aug 1815.

Rogue, Mrs. - 7th U.S. Infantry - Order book, 30 Jan 1814 to 5 Feb 1815, present.

Rollings, Mary - 4th U.S. Infantry - Company: Captain Ebenezer Way - Admitted to hospital, 7 Sep 1814 and discharged from hospital, 30 Sep 1814.

Romley, Eliza - Seaman – U.S. Navy - Captured on board the U.S. Sloop Growler on Lake Champlain on 3 Jun 1813 disguised at a man; interned at Quebec, Canada, and released twenty-two days later.

Russell, Sarah - 32nd U.S. Infantry - Company: Captain Jonathan Roberson - Enlistment date: 15 Aug 1813 - Monthly returns, 1 Nov 1813, present, on rations.

The men, women and children

Sands, Sarah - Nurse - Hospital Attendants, Pennsylvania.

Skinkle, Colly - Washerwoman - 9th Regiment, New York Militia (Lieutenant Colonel Peter L. Vosburgh).

Sparks, Elizabeth - Washerwoman - 42nd U.S. Infantry - Company: Captain Jonathan Roberson - Inspection return, Philadelphia, PA, 28 Feb 1815.

Stone, Mary - Civilian - 1st U.S. Infantry - Prisoner of War at Quebec; wife of Sergeant John Stone.

Stoner, John - Child - 1st U.S. Infantry - Age: 2 months - Died at Quebec, Canada, on 6 Oct 1812 while a prisoner of war; parents: Sergeant John and Mary Stoner.

Stultz, Fanny - Washerwoman - 19th U.S. Infantry - Company: Captain Wilson Elliott - Wife of Adam Stultz.

Suit, Susanna - 14th U.S. Infantry - Admitted to hospital, 7 Sep 1814 and discharged from hospital, 30 Sep 1814.

Swan, Rachael - Servant – Brigadier General Martin Herrmance's Brigade, New York Militia.

Swinburn, Johanna P. - Nurse - Hospital Attendants, Pennsylvania.

Tayler, Mrs. - Washerwoman - 8th Regiment, New York Militia (Lieutenant Colonel Thomas Miller).

Todd, Rebecca - Washerwoman – U.S. Sea Fencibles - Fort Constitution, MA; absent Feb 1815.

Trask, ----- - Washerwoman - Woman - Fort Wissasset, ME, Feb 1815.

Tucker, Mary - Washerwoman – U.S. Sea Fencibles.

Van Dyne, Isabella - Nurse & acting washerwomen - 3rd U.S. Military District - Monthly return, Hospital Bellevue, 30 Sep 1814; nurse and acting washerwoman; monthly return, 30 Jan 1815; discharged 26 Jan 1815.

Walker, Jane - Servant - Major General William Butler's Division, South Carolina Militia.

Walker, Mrs. - 5th U.S. Infantry - Company: Third Lieutenant Edward Upham's Detachment - Roll dated 28 Feb 1815.

Waters, Polly - Washerwoman - 19th U.S. Infantry.

Weir, Ann - Civilian - 4th U.S. Infantry - Prisoner of war at Quebec, Canada; wife of Private David Weir.

Weir, Lucinda - Child - 4th U.S. Infantry - Age: 13 months - Died at Quebec, Canada, on 14 Oct 1812 while a prisoner of war; parents: Private David and Ann Weir.

Wentling, Ann - Hospital Nurse - 1st Regiment, Riflemen, Pennsylvania Militia (Colonel Thomas Humphrey).

Whitelock, Jane - Child - 1st U.S. Infantry - Age: 2 years - Died at Quebec, Canada, on 12 Oct 1812 while a prisoner of war; father: Sergeant John Whitelock.

Wickam, Maryhew - Waiter - 3rd U.S. Military District - Hospital Bellevue, Sep 1814.

Wingard, Hannah - Matron - Fort Niagara, NY - Company: Hospital - Medical staff at Niagara, Dec 1814.

Woodruff, Rhoda - Servant – U.S. Corps of Engineers - Enlistment date: 1 Jun 1814 - Monthly return, U.S. Military Academy, West Point, NY, 30 Sep 1814; private servant to Dr. Samuel A. Walsh, Post Surgeon, Feb-Apr 1815, present.

Yeaton, Abigail - Washerwoman - 40th U.S. Infantry - Company: Captain Mathew Sanborn - Fort Constitution, MA; Feb 1814-Dec 1815.

Youmans, Catherine - Servant – U.S. Corps of Engineers - Enlistment date: 1 Jun 1814 - Monthly return, U.S. Military Academy, West Point, NY; private servant to Captain Alden Partridge, Professor Art of Engineering, Apr 1815.

Young, Mary - Hospital Matron - 42nd U.S. Infantry - Company: Captain William Oliver - Monthly return, Province Island Barracks, PA, Oct 1813, present; hospital matron entitled to pay.

## Bibliography

*American soldiers and their children who died in Quebec city during the War of 1812,* Quebec Weekly Chronicle, Thursday, 18 July 1889, Volume 42, Number 3,656, Page 2, Columns 3, 4 and 5.

Bromley, Viola A., *The Bromley genealogy: being a record of the descendants of Luke Bromley of Warwick, R. I., and Stonington, Conn., (Frederick H. Hitchcock, Genealogical Publisher: New York, NY 1911), page 43, Eliza Bromley.*

"Compete muster roll of a detachment of recruits," 9 August 1812, *George Tod Papers 1783-1834*, Western Reserve Historical Society Archives Library, Cleveland, Ohio, manuscript section, call number MS-3202, container 2a.

"Compete muster roll of a detachment of recruits," 18 September 1812, *George Tod Papers 1783-1834*, Western Reserve Historical Society Archives Library, Cleveland, Ohio, manuscript section, call number MS-3202, container 2a.

*History of Allegany County, New York, 1806-1879,* (F. W. Beers & Co., NY; 1879), page 319.

"List of artillery men sent to Chillicothe," 29 October 1812, *George Tod Papers 1783-1834*, Western Reserve Historical Society Archives Library, Cleveland, Ohio, manuscript section, call number MS-3202, container 2a

Fairchild, Jr., C. M., *Journal of an American Prisoner at Fort Malden and Quebec in the War of 1812*, (Frank Carrel; Quebec, Canada, 1909).

*General Entry Book of American Prisoners of War* ledger of the British Admiralty made by the Public Record Office in London, Great Britain (ADM 103 / 362).

Johnson, Eric Eugene, *American Prisoners of War held at Quebec during the War of 1812, 8 June 1813 – 11 December 1814*, (Heritage Books, Inc.: Westminster, MD 2011), pages iv and 132, Eliza Romley.

Lindly, Harlow, *Fort Meigs and the War of 1812*, (The Ohio Historical Society, Columbus, Ohio: 1975).

*Miscellaneous Lists and Material of the British Admiralty* housed at the Public Record Office in London, Great Britain (ADM 103 / 465, part 2).

*Public Statutes at Large of the United States of America*, Volume II, (Boston: Charles C. Little and James Brown, 1845), Seventh Congress, Session I, Chapter IX, pp. 132-137, 16 March 1802, "An act fixing the military peace establishment of the United States."

*Public Statutes at Large of the United States of America*, Volume II, (Boston: Charles C. Little and James Brown, 1845), Twelfth Congress, Session I, Chapter CXXXVII, pp. 784-785, 6 Jul 1812, "An act making further provision for the Army of the United States, and for other purposes."

*Records Relating to War of 1812 Prisoners of War, 1812*; (National Archives Microfilm Publication M2019, 1 roll); Records of the Adjutant General's Office, 1780's-1917; Record Group 94, number 68, pages 1-4; National Archives, Washington, D.C.

*Register of Enlistments in the U.S. Army, 1798-1914*; (National Archives Microfilm Publication M233, 81 rolls); Records of the Adjutant General's Office, 1780's-1917, Record Group 94; National Archives, Washington, D.C.

*War of 1812 Pension Applications.* Washington D.C.: National Archives. NARA Microfilm Publication M313, 102 rolls. Records of the Department of Veterans Affairs, Record Group Number 15, Old War Invalid Rejected File 12502, Joseph Bledsoe.

## Jeremiah Buchanan and his three sons

How many men do you think ever had the opportunity of enlisting in the U.S. Army along side their sons? Jeremiah Buchanan of Roane County, Tennessee, had his chance during the War of 1812 when on 24 December 1812 he enlisted with three of his sons in the 24th Regiment of U.S. Infantry.

What is unique is that all of his sons were under age. Hercules Buchanan was 16 years of age while his brother Moses was 14 and another brother Jeremiah was 12. The boys were not musicians. They signed on for a five-year enlistment as soldiers.

The 24th Infantry was raised in Tennessee and in the Territory of Mississippi. Three companies served with Major General Andrew Jackson in the Army of the Southwest while two companies served under Brigadier General Benjamin Howard at St. Louis. The remaining five companies were transferred to the Army of the Northwest to serve under Major General William Henry Harrison.

In June 1813, the five companies arrived in Ohio and they were stationed at Fort Stephenson on the Sandusky River where Fremont now stands. The regiment was reassigned to Fort Meigs on the Maumee River, just south of present-day Toledo, in late August. The regiment was attached to Brigadier General Duncan McArthur's brigade which was preparing for the invasion of Upper Canada. The 17th Infantry from Kentucky and the 19th Infantry from Ohio were the other two regiments assigned to this brigade.

The 24th Infantry participated in General Harrison's invasion of Upper Canada in late September but it was not involved in the Battle of the Thames River in October. Two companies from the 24th Infantry were transferred to the Niagara Theatre along with half of the Army of the Northwest.

In late October the Buchanan's found themselves in Buffalo, New York. Jeremiah Buchanan, the father, took sick and he was left behind on 3 November 1813 when the Army of the Northwest was ordered to Fort George on the Canadian side of the Niagara River. He died in the Buffalo area in December.

The Army of the Northwest was then ordered to Sackets Harbor, New York, where they remained until the end of the war. The three sons, in the meantime, had transferred to the Regiment of U.S. Rifles at this facility. The regiment would receive its "1st" designation the following year when three additional rifle regiments were formed.

Due to their enlistments, the boys remained in the army for another two years. By their service records it appears that they were separated. Jeremiah, the son, was transferred to Belle Fontaine, Missouri Territory, on 31 December 1815. He was stationed at Prairie de Chien on 30 June 1816 and he was at Fort Edwards when he was discharged from the army on 24 December 1817.

Moses was transferred to Fort Mackinaw, Michigan Territory, where he served from 30 September 1815 to 31 October 1817. He was then transferred to Fort Crawford where he was discharged on 24 December 1817.

The service record of Hercules Buchanan does not state where he was stationed after the war. On 24 December 1817 the three sons, now men, headed home from different locations without their father. But this does not end this story.

Another son, Thomas, had enlisted in the same rifle regiment on 24 September 1811. He was at Buffalo with his father and brothers and he too sat out the war at Sackets Harbor. After the war, Thomas was transferred to the south where he was stationed at Fort Hawkins, Georgia, and later at Fort Montgomery and Fort Crawford in what is now Alabama. He was discharge on 24 September 1816.

Although he died during the war, Jeremiah Buchanan had the unique opportunity to serve with four of his sons while he was stationed at Buffalo. It is not the generals who win the wars; it is the supreme sacrifice that a family makes in sending their loved ones off to war.

## The Mystery of Captain Benjamin LeRoy

A friend found an interesting auction item on the Internet of a photograph of a veteran of the War of 1812 who claims to have fought at the Battle of Lundy's Lane. The seller described the lot as, "This is an interesting CDV[1] image of Captain Benjamin LeRoy who claimed to be a veteran of the War of 1812 and the battle of Lundy's Lane. The image is of an old man with long white hair and beard. He appears to be wearing a Civil War Era Federal sack coat and has a large pin or medal on his vest. Since the card is from Dayton I expect that it was made while he was a resident of the soldier's home there."

Benjamin LeRoy

Finding a photograph of a common soldier of the War of 1812 is an exciting discovery. Most of these men were photographed in the twilight of their years. Many men, particularly officers, who had successful careers after the war, had their portraits painted or photographs taken. For most Americans having a photograph made was an expensive undertaking especially before the Civil War.

Now to unravel the mystery of who was Captain Benjamin LeRoy.

The Central Branch of the National Asylum for Disabled Volunteer Soldiers was founded in 1867 in Dayton, Ohio. The Asylum is commonly called the Dayton Soldiers Home. Records from this institute show that an army veteran, Benjamin F. LeRoy, died at the home on 6 October 1878. He was born in 1780.

The pension listings for this war also have a Benjamin LeRoy, whose widow, Mariann, applied for a survivor's pension, number SC-269. This LeRoy served as a private in Captain Lucke Parson's Company of the Vermont Militia. The National Archives has only one service record for a Benjamin LeRoy and he served in the 3rd Regiment of the Vermont Militia. Lieutenant Colonel William Williams was probably the regimental commander.

Checking U.S. Army records, there are no officers listed in the official officer's register for the war with the name of Benjamin LeRoy. Also, the only men from Vermont who fought at the Battle of Lundy's Lane on 25 July 1814 were from the 11th Regiment of U.S. Infantry. There were no militia units from Vermont participating in this battle.

Two letters to the Dayton Soldiers Home requesting information on Benjamin LeRoy were not answered. The next step was to obtain a copy of his wife's widow pension.

The pension package lists two survivor's numbers, SO-537 and SC-269. It appears that Benjamin had applied for an invalid pension but apparently this was rejected. His wife would receive the widow's pension of eight dollars per year for a husband who served as a private. Benjamin had been issued a land bounty during the 1850's for his service in the militia during the war.

In Benjamin's Declaration for Pension he states that he served a full 60 days in Captain Luke Parson's Company of Horse, 1st Vermont Regiment, from 30 August 1812 to 1 January 1813 (total of four months). This company patrolled the border between Vermont and Canada. He and his wife were living in Detroit, Michigan, before he was admitted to the Dayton Soldiers Home.

A letter from the home, which was included in the pension papers, states that Benjamin died at the home's hospital on 6 October 1878. Private Benjamin F. LeRoy was buried in the Dayton National Cemetery in Dayton, Ohio. He was laid to rest in section A, grave 25, row 14.

The records from the Dayton Soldiers Home does state that Benjamin did served in the 11th Regiment of U.S. Infantry. He was not an army officer although he may have been a militia captain after the war. The Vermont militia records would have to be researched to verify that he was a militia officer.

If Benjamin had been a captain then this information should have been mentioned in the pension papers. Benjamin LeRoy was a simple soldier serving his country during the War of 1812. It may never be known why he is labeled as a captain and but he was a veteran of this famous battle.

---

[1] Carte de visite (CDV) or visiting card image.

## Midshipman John L. Cummings: A Case Study

John L. Cummings was born in New Jersey and he was appointed as a midshipman in the U.S. Navy on 8 October 1800. Most likely, he was a 'boy', a naval rank, before being promoted to midshipman. He may have participated in the Quasi-War with France between 1798 and 1800. After his promotion, he probably also participated in the First Barbary Wars between 1801-1805. Further research will need to be conducted to prove these two theories.

Midshipman Cummings was assigned to the U.S. Frigate *Constitution* after the ship had captured the H.M.S. *Guerriere* on 12 August 1812. He spent a short time with the *Constitution* before being reassigned to the Great Lakes along with three other men from his ship.

On the night of 8 October 1812 Cummings was in a party of American soldiers and sailors under the command of Lieutenant Jesse D. Elliott, USN, when they captured from the British the former U.S. Army Brig *Detroit* and the Provincial Marine Ship *Caledonia* at the head of the Niagara River near Buffalo, New York. The *Detroit* was destroyed while the *Caledonia* became the U.S. Brig *Caledonia*. Cummings was wounded during this engagement. During the Battle of Lake Erie on 10 September 1813, Cummings was serving on the U.S. Brig *Niagara* when he received his second wound.

Cummings was captured by the British on the night of 12 August 1814 while serving on the U.S. Schooner *Ohio*. Both the *Ohio* and the U.S. Schooner *Somers* fell into British hands while the ships were supporting the U.S. Army operations with naval bombardments during the Battle of Fort Erie in Upper Canada.

*American Prisoners of War held at Quebec during the War of 1812* shows that Cummings was prisoner number 1566 with the rank of midshipman who had been captured on 12 August 1814 while serving on the gunboat *Ohio* off Fort Erie. He arrived at Quebec with other prisoners by way of a steamboat on 16 September 1814. He was discharged from Quebec on 10 November 1814 and sent to Halifax on H.M. Transport *Lord Cathcart*. Cummings' name appears on the roster of American POWs being sent to Halifax as listed in the *American Prisoners of War held in Montreal and Quebec during the War of 1812*.

*American Prisoners of War held at Halifax during the War of 1812* shows that Cummings arrived at Halifax on 6 December 1813 on the H.M. Transport *George*. He was discharged on 14 March 1815 and sent to Salem, Massachusetts on the H.M. Transport *Hope 3*. Apparently, new orders had been received and the ledger was not updated since Cummings' name next appears on a ledger in England. While at Halifax, he was prisoner number 7962.

Cummings arrived in England on the H.M. Transport *George* and he probably went through the POW processing center at Mill Prison in Plymouth. He was paroled at Dartmouth on 6 December 1814 and remained at this city until 14 March 1815 when he was released and sent home on the H.M. Transport *Hope 3* according to the *American Prisoners of War Paroled at Dartmouth, Halifax, Jamaica and Odiham during the War of 1812*. While paroled at Dartmouth, Cummings' prisoner number was still 7962.

After the war, Cummings was promoted to lieutenant on 1 April 1818 while assigned to the Philadelphia Naval Yard. He died on 24 July 1824 while still in the navy. There has been no U.S. naval ships named for Lieutenant John L. Cummings.

## Richard Sparks: Indian fighter, explorer, soldier

Few men can match the accomplishments that Richard Sparks achieved prior to and during the War of 1812. As a career army officer, Sparks made his mark on American history but he is not remembered today.

Richard was born in New Jersey around 1757 and as a young boy, he was captured by Shawnee Indians after his family had moved to western Pennsylvania. He became the adopted brother of Tecumseh and was given the Indian name "Shawtunte." He was re-united with his white family after Lord Dunmore's War under the 1775 Treaty of Camp Charlotte.

During the Revolutionary War, he served as a sergeant for a short time in the 9th Virginia Regiment (Continental Line), he was a captain in the Pennsylvania militia, and he served as a scout during Colonel William Crawford's Expedition of 1782 into the Ohio region.

Sparks was commissioned a captain in the U.S. Levies in 1791 and he was a member of Major General Arthur St. Clair's army. He was on a scouting mission when St. Clair's army was defeated by Indians on 4 November 1791 near present-day Fort Recovery, Ohio.

On 7 March 1792 he received a captain's commission in the regular army. He was at the Battle of Fallen Timbers with Major General Anthony Wayne on 20 August 1794 near present-day Toledo, Ohio.

After the Indian wars in the Old Northwest Territory, Sparks remained in the U.S. Army. He was transferred to the 3rd Regiment of U.S. Infantry on 1 November 1796 and then the 2nd Regiment of U.S. Infantry on 1 April 1802.

Colonel Richard Sparks

He was tasked by President Thomas Jefferson to lead one of the three exploring expeditions to determine the size of the newly purchased Louisiana Territory. Captain Meriwether Lewis and William Clark's (Lewis and Clark Expedition 1804-1806) goal was to find the source of the Missouri River which would established the western boundaries of the new territory. Captain Zebulon Pike, Jr. (Pike Expedition 1805) would determine the source of the Mississippi River which would set the northern boundaries of the territory. Finally, Sparks would lead the Red River Expedition of 1806. This party's goal was to establish the southern boundaries of the territory.

Sparks' expedition failed when he and his party were arrested by Spanish troops after they had entered present-day Oklahoma. They were released and forced to return to New Orleans. Captain Pike was then tasked to finish Sparks' mission. Pike's second expedition was a success and the tallest mountain that was seen by Pike's party in Colorado was named after him, Pike's Peak.

Captain Sparks returned to the 2nd Infantry and was promoted to major on 29 July 1806, to lieutenant colonel on 9 December 1807 and then to colonel on 6 July 1812. He was the commander of the 2nd Infantry throughout the War of 1812.

The regiment was headquartered in New Orleans and the regiment was a unit of Major General Andrew Jackson's Army of the Southwest. Part of the regiment participated in the First Battle of Fort Boyer on 15 September 1814 and the Second Battle of Fort Boyer on 11 February 1815. The fort protected the entrance to Mobile Bay in Alabama.

Sparks suffered a stroke and was forced to resign from the army on 15 June 1815. He died shortly afterwards on 1 July 1815 in Claiborne County, Mississippi.

Richard Sparks' military career stretched from the beginning of the Revolutionary War to the end of the War of 1812. He served in the militia, the levies and the regular army as a soldier, as an officer, as an explorer and as an Indian fighter.

## Edmond Forbes Duvall: An Ohioan at the Battle of Baltimore

Edmond Forbes Duvall is an unique Ohio veteran from the War of 1812 having served with Brigadier General Edward W. Tupper on the Ohio frontier in 1812 fighting Indians and then traveling to Maryland to join Commodore Joshua Barry's Flotilla. Duvall participated in the Battle of the Maumee Rapids (Ohio) on 14 November 1812, the Battle of Bladensburg on 24 August 1814, and the Battle of Baltimore on 13 September 1814.

Duvall was born on 23 May 1794 to Jacob Duvall and Jemima Anne Taylor in Prince George's County, Maryland. His father was a lieutenant in the Maryland militia during the Revolutionary War. Jacob died in 1800 and his mother married a relative of Jacob's in 1804 named Charles Duvall.

Charles moved his family to Muskingum County, Ohio, shortly after his marriage and he operated either a keel boat or a flat-bottom boat on the Muskingum, Ohio, and Mississippi Rivers. He would transport grain and probably spirits between Zanesville and Marietta in Ohio and then to New Orleans, and return with goods needed for the frontier. Edmond probably accompanied his step-father on these inland voyages where he gained the small boat experience he needed as a member of Barry's Flotilla.

In 1805 while at New Orleans, Jacob Duvall met Aaron Burr, the third vice-president of the United States and the man who killed Alexander Hamilton in an 1804 duel. Charles met Burr again in September 1806 at Marietta where Burr tired to recruit Jacob for his private army. Burr had plans to take over parts of the newly acquired Louisiana Territory and parts of the territory held by Spain in order to create his own country. Burr was caught and charged with treason before his could carry out his plans. Duvall testified against Burr on 24 September 1807 in Burr's conspiracy trial which was held in Richmond, Virginia. Burr was acquitted.

With the outbreak of war in 1812, Edmond Duvall joined Captain James Wimp's Company of the Ohio militia at Zanesville, and he served between June 1812 and 20 December 1812. Wimp's company was a part of General Tupper's brigade which was the central division of Brigadier General William Henry Harrison's Army of the Northwest. Tupper's orders were to march north from Urbana, Ohio, and secure the Maumee Rapids, near present day Toledo. Fort Meigs would later be built near the rapids. Tupper's forces were attacked by Indians on 14 November 1812 and the Indians were driven off. With dwindling supplies, the brigade was forced to retreat back to Urbana.

In 1814, Edmund secured a commission in the U.S. Flotilla Service as a midshipman and joined Captain Joshua Barney's Chesapeake Bay Flotilla at Baltimore, Maryland. He served from 1 March 1814 to 1 April 1815. Barney's rank was 'captain' but he used the title 'commodore' since he commanded more than one vessel. The rank of commodore would not be created by the U.S. Congress until 1862.

While a member of the flotilla, Duvall participated in the Battle of St. Jerome Creek (1 Jun 1814), Battle of St. Leonard's Creek (26 Jun 1814), the Battle of Queen Anne (22 Aug 1814), the Battle of Bladensburg and the Battle of Baltimore. He served on Lieutenant Solomon Rutter's gunboat during the first three engagements. During the Battle of Baltimore, which included the attack on Fort McHenry, Duvall either served at the Lazaretto Battery, which was across the Patapsco River from Fort McHenry, or he commanded one of the fourteen gun barges on the river. After the battle, he was stationed at Fort McHenry until discharged.

Edmond returned to Muskingum County after the war. He lived for a short time in Fulton County, Illinois, during the 1850s and then returned back to Ohio. He died on 19 May 1881 at the Muskingum County Infirmary in Zanesville and he is buried in the infirmary's cemetery. He never married. He received 160 acres of military bound lands in Barton County, Missouri, which he sold to John Postlewait on 1 July 1859. He received a military pension in 1871 in which he outlined his military service.

On the 200th Anniversary of the Battle of Baltimore, Paul Morehouse, Eric Johnson, Richard Davis, Eric Leininger and Craig Fisher, members of the Society of the War of 1812 in the State of Ohio, witnessed the re-enactment of the Battle of Fort McHenry and the raising of Old Glory. At that time, we did not realized that another Ohioan, Edmond Forbes Duvall, had witnessed the real Battle of Fort McHenry and the raising of the original Old Glory.

## Bibliography

*American State Papers, Documents, Legislative and Executive of the Congress of the United States, volume 1 – 1789-1809, Miscellaneous*, (Washington, D.C.: Gales and Seaton, 1832), Burr's Conspiracy Trial at Richmond, Virginia, 22 May 1807, pp. 533-535.

Edmond F. Duvall Pension, number SO-2629 and SC-5769, War of 1812, War Department Collection, National Archives & Records Administration, Washington, D.C

Military land bounty, John Postlewait, assignee of Edmund F. Duvall, Scrip Warrant Act of 3 March 1855, dated 1 July 1859, warrant number 58138, 160 acres, General Land Office Records, Bureau of Land Management, U.S. Department of Interior, Washington, DC.

Muskingum County, Ohio, Death Records 1867-1902, pp 52, Zanesville, Ohio.

Newman, Harry Wright, *Mareen Duvall of Middle Plantation*, (Washington, DC 1952), pp. 261-262.

*Roster of Ohio Soldiers in the War of 1812*, The Adjutant General of Ohio 1816, (Heritage Books, Inc., Bowie, Maryland: 1995), Captain James Wimp's Company, page 69.

# Ohio's causalities during the Battle of Chippewa

Three Ohioans are listed on the causality report for the Battle of Chippewa as having been killed in action on 5 July 1814 during the War of 1812. The battle was fought on the Canadian side of the Niagara River, a few miles south of Niagara Falls.

The soldiers were members of the 19th Regiment of U.S. Infantry, which was organized in Ohio. Part of this regiment was transferred to the state of New York after the Battle of the Thames River in late 1813 to support U.S. Army operations in the northern theatre. The names of these three men, as listed on a muster roll taken after the battle, were Privates Joseph Mingro, James Mu-----, and Elijah Mullenic. James' last name is unrecognizable.

However, the *Register of Enlistments in the U.S. Army, 1798-1914* shows that names of these men were Elijah Mullinax, James Mullinax, and Joseph Mengro. Further evidence will indicate that Elijah and James were brothers while Joseph was not killed in battle but became an American prisoner of war in Canada.

Muster rolls were required after a battle in order to account for each man in a regiment to determine who were still fit for duty and who were causalities. A causality is any man who was killed, wounded, died from wounds, captured by the enemy or who had deserted. A company clerk or an officer who filled out these musters may not always know what happened to each man in his company. In fact, it may be months before everyone was accounted for. This was the case for the three men from Ohio.

The Mullinax brothers were recruited by First Lieutenant Charles Lee Cass of the 27th Regiment of U.S. Infantry in November 1813 in the Newark-Zanesville area of Ohio. Lieutenant Cass was a recruiting officer for this regiment, and his recruiting reports for August through September 1813 shows him recruiting in Zanesville, while the next surviving report shows him recruiting in Newark during February 1814.

The 27th Infantry, along with the 26th Infantry (both of these regiments were also raised in Ohio), was disbanded on 12 May 1814 and merged with the 26th Infantry to reform the 19th Infantry. The original 19th Infantry was disbanded and its soldiers were assigned to the reformed 17th Infantry.

John Mullinax, the brother to Elijah and James, became the principle heir of his deceased brothers. Bounty land warrant number 22,311 was issued to John for his brother James, while number 22,312 was issued to John for Elijah. The first warrant states, "John Mullinax, brother and other heirs at law of Elijah Mullinax, deceased," while the second warrant indicates James Mullinax's name. James' 160 acres of land was located in Lonoke County, Arkansas, while Elijah's 160 acres was located in Arkansas County, Arkansas. John received patents for both properties in 1820.

Joseph Mengro enlisted in the 19th Infantry in March 1814. His recruiting officer was Third Lieutenant David L. Carney. Lieutenant Carney's recruiting reports state that he was recruiting in Chillicothe, Ohio, between February and April 1814.

According to the *General Entry Book of American Prisoners of War* ledgers of the British Admiralty, Mengro was a prisoner at both the Quebec and Halifax prisoner of war facilities. He was captured on 5 July 1814 and arrived at Quebec on 11 August 1814 aboard the Transport Royal Seaman. Mengro's name was recorded as Joseph Minger and his prisoner number was 1,469. He was discharged on 8 October 1814 and sent to Halifax on the Transport Queen.

The Halifax prison ledger states that his name was Joseph Migno and he was given the prisoner number 7,704. He arrived at the prison on the Transport Queen on 31 October 1814. He was discharged at the war's end on 10 April 1815 and he was sent home to the United States on the Transport Jubilee. This ledger maintains that he was captured by Canadians on 2 September 1814.

According to Mengro's service records, he arrived at Fort Independence, Boston, Massachusetts, on 20 April 1815 on the Cartel Brig Jubilee. The record states that he was a prisoner of war from Halifax and his name was recorded on the muster roll from this fort on 26 April 1815. A 'cartel' ship flew the white flag and they were used by warring countries to exchange prisoners of war. They were all used to transport official documents and government representatives.

When Joseph Mengro was discharged from the army is not recorded in this service records. However, he had enlisted for the duration of the war so he would have been discharged after recovering from his ordeal as a prisoner of war. He is not listed after the war as having received either a bounty land or a pension.

## Bibliography

Baker, Harrison Scott II, *American Prisoners of War Held at Halifax During the War of 1812*, (Heritage Books, Inc.: Westminster, MD, 2004), volume 1, page 275.

Graves, Donald E., *Red Coats & Grey Jackets: The Battle of Chippawa, 5 July 1814*, (Dundurn Press Limited, Toronto, Canada: 1994), Appendices, American Regulars, Militia and Native Warriors killed at Chippawa, page 173.

Heitman, Francis B., *Historical Register and Dictionary of the United States Army From Its Organization, September 29, 1789, to March 2, 1903*, Volume I, (Genealogical Publishing Company, Baltimore, Maryland: 1994).

Johnson, Eric Eugene, *American Prisoners of War Held at Quebec During the War of 1812*, (Heritage Books, Inc.: Westminster, MD, 2011), page 106.

*Register of Enlistments in the U.S. Army, 1798-1914*; Records of the Adjutant General's Office, 1780's-1917, Record Group 94; National Archives, Washington, D.C.

United States. Bureau of Land Management, General Land Office Records. Automated Records Project; *Federal Land Patents*, State Volumes: http://www.glorecords.blm.gov/

*War of 1812 Military Bounty Land Warrants, 1815-1858*; National Archives Microfilm Publication M848; Records of the Bureau of Land Management, Record Group 49; National Archives, Washington, D.C.

*War of 1812 Pension Applications*; National Archives Microfilm Publication M313; Records of the Department of Veterans Affairs, Record Group Number 15; National Archives, Washington D.C.

# The Grinton Brothers of North Carolina

It is not unusual for brothers to enlist in the U.S. Army in times of war but it is unusual that three African American brothers served together during the War of 1812 in the 10th Regiment of U.S. Infantry. The army's policy during this war was that Blacks could not serve as soldiers, only as officer's servants or in a non-combatant role.

Since each regiment had its own recruiting service during this war, many colonels over looked the army's ruling and had their recruiting officers enlist Blacks, especially light-skinned Mulattoes. Approximately 400 Blacks were recruited by the army while another 800 served in the Louisiana Militia. Louisiana was the only state that permitted free Blacks to serve in its militia forces. It may be never known how many light-skinned Mulattoes served in the land forces of the United States during the War of 1812.

The 10th Regiment of U.S. Infantry was authorized on 11 January 1812 and North Carolina was tasked with raising this regiment. The regiment was disbanded on 17 May 1815 and most of the men who had time remaining on their enlistments were transferred to the new 6th Regiment of U.S. Infantry.

Most of the regiment served with the Army of the North in New York and participated in the Battles of Chateauguay and Lacolle Mill, both in present day Quebec. The Grintons served in a company that was stationed at Fort Washington, Maryland, just south of Washington, DC.

The Grinton family is listed in the U.S. census records starting in 1790 as free Blacks living in Wilkes County, North Carolina. Starting in 1850 they are listed either as Blacks or Mulattoes, still free.

Martin, Philip and Robert Grinton were all recruited by 3rd Lieutenant Wilie Gordon of the 10th Infantry in April 1813 in Wilkes County. Martin was 20 years old and listed as 'Colored' on the army's Register of Enlistments while Philip was also 20. He is not listed as 'Colored' and his physical description could pass as White. Robert is 24 and is also listed as 'Colored.'

All three men served in Captain Josiah Wood's Company (later, Captain Emanuel Leight's Company) at Fort Washington, Maryland, according to Martin's pension application. Philip was transferred to the 8th Regiment of U.S. Infantry at the war's end and he died on 26 September 1815 from ague and fever at Portage de Sioux, Missouri, while still serving in the army. Martin is listed as the brother of Philip when he received his dead brother's military land bounty on 14 March 1825. The land was in Arkansas.

Robert was hospitalized on 7 September 1814 and he returned to duty on 31 October 1814. He was placed on recruiting duties but he was discharged as unfit on 25 August 1815. He has not been found after this date. Robert, as the older brother of Philip, should have received his brother's land bounty along with the rest of the legal heirs. He may have returned home and died there before 1825.

Martin made the ultimate mistake in the army. He deserted on 22 July 1815, was captured on 25 July 1815, court-martialed and then discharged on 27 August 1815. He forfeited all back pay, bonuses, bounties and pensions. He did apply for a pension on 2 November 1872 but his application was rejected on 2 November 1872 due to his desertion. He was living in Americus, Sumter County, Georgia, when he applied for a pension.

Martin made the mistake that many young men did at this time. He deserted, not because he was a coward, but because the war had ended, he had done his duty, he was going home. He paid a heavy price since he still had time to serve on his enlistment.

## Bibliography

Heitman, Francis B., *Historical Register and Dictionary of the United States Army From Its Organization, September 29, 1789, to March 2, 1903*, Volume I, (Genealogical Publishing Company, Baltimore, Maryland: 1994).

*Register of Enlistments in the U.S. Army, 1798-1914*; (Washington, DC: National Archives Microfilm Publication M233); Records of the Adjutant General's Office, 1780's-1917, Record Group 94.

Bureau of Land Management, General Land Office Records, *Federal Land Patents*, State Volumes, http://www.glorecords.blm.gov/, Springfield, Virginia.

*War of 1812 Military Bounty Land Warrants, 1815-1858*; (Washington, DC: National Archives Microfilm Publication M848); Records of the Bureau of Land Management, Record Group 49.

*War of 1812 Pension Applications*; (Washington D.C.: National Archives, NARA Microfilm Publication M313), Records of the Department of Veterans Affairs, Record Group Number 15.

# Samuel Orwick

Samuel Orwick came to Ohio from Pennsylvania as a teenager with his mother, Mary, his step-father Philip Carrel (Carroll) and nine half-siblings. The name of his father and his mother's maiden name is not known at this time. The family settled in Knox Township, Jefferson County, prior to 1813. The Carrel family would eventually use the Carroll surname.

Orwick served in the regular army during the War of 1812 with the 17th Regiment of U.S. Infantry. This regiment recruited in both Kentucky and Ohio. Samuel's military service papers recorded that he served as a private.[2] He was 5' 5" tall with blue eyes, dark hair and dark complexion. He was a farmer who had been born in Bucks County, Pennsylvania. He enlisted at 18 years of age on 2 November 1813 to serve for the duration of the war.

He was enlisted by Third Lieutenant Isaac M. Rieley of the 7th Regiment of U.S. Infantry in Springfield Township, Jefferson County. Lieutenant Rieley had been recruiting without permission in Ohio and when he sent his recruits to the recruiting center at Chillicothe, Ohio, for processing, they were re-assigned to the 17th Infantry. Had the lieutenant taken his recruits out of Ohio himself, Samuel could have taken part in the Battle of New Orleans in January 1815. Most of the 7th Infantry was assigned to New Orleans during this war.

At Chillicothe, Orwick was assigned to Captain Harris H. Hickman's company which was stationed at Erie, Pennsylvania, during the winter of 1814-1815. This company was a part of an army detachment which was protecting the Lake Erie naval squadron while the squadron was in winter quarters. Upon the end of his enlistment term, he was assigned to Third Lieutenant William Featherston's discharge detachment and was discharged from the army at Chillicothe, Ohio, on 9 June 1815.

Samuel received 160 acres of land in Illinois for his service in the war but the warrant was sent to Samuel Smith of Union County, Pennsylvania.[3] Samuel waited until 23 March 1839 when he wrote to his Congressmen requesting a new warrant. His first request was rejected on 31 August 1839 since Smith had received a patent (deed) for this land.

It would not be until 1845 before Samuel would be successful in obtaining a new land bounty warrant. This new warrant gave him 160 acres of land in Bureau County, Illinois. Samuel obtained a deed for his property but never settled on this land nor did he ever pay taxes on his property. His land was sold at a sheriff's sale.

Samuel's step-father, Philip Carrel, served a tour of duty with Captain Robert Gilmore's Company from Jefferson County between 10 February 1813 until 25 May 1813 when Philip enlisted in the U.S. Army while his company was stationed at Cleveland, Ohio.[4] He served in Captain Stanton Sholes'

---

[2] *Register of Enlistments in the U.S. Army, 1798-1914*; Samuel Orwick, 17th Regiment of U.S. Infantry, National Archives Microfilm Publication M233; Records of the Adjutant General's Office, 1780's-1917, Record Group 94; National Archives, Washington, D.C.

[3] *War of 1812 Military Bounty Land Warrants, 1815-1858*; warrant number 16,438, issued to Samuel Orwick, National Archives Microfilm Publication M848, Records of the Bureau of Land Management, Record Group 49; National Archives, Washington, D.C.

[4] *Roster of Ohio Soldiers in the War of 1812*, (Adjutant General of Ohio: Columbus, OH 1916), page 97, Captain Robert Gilmore's Company, Private Phillip Carroll.

Company from the 2nd Regiment of U.S. Artillery at Fort Huntington in Cleveland.[5] This company participated in Major General William Henry Harrison's invasion of Canada in late September 1813 and was then assigned to Fort Shelby (formerly Fort Detroit) in Detroit, Michigan. An epidemic broke out at the fort in early December which lasted into the winters months of 1814. Philip died from disease on 10 December 1813 while at Fort Shelby.

Samuel came home from the army and found that his mother's family had been broken up with many of this half brothers and sisters living with probate court assigned guardians. He would remain close to his mother and half-sister Margaret. No land bounty has been found for the heirs of Philip Carrel, so most likely, Mary accepted Philip's pay (half pay) for five years as her pension. Her children, by Carrel, would receive pensions until their 16th birthday.[6] The *Pension List of 1835* lists the children as John, Mary, Margaret, Joseph, Armstrong (sic), Catherine, Henry, Jane and Philip Carroll. Armstrong is actually Anthony Wayne Carroll.

Samuel married Sarah Palmer, the daughter of Joseph Palmer and Sophia Oldfield, on 11 November 1817 in Columbiana County, Ohio.[7] Sarah went by the nickname of "Sally" as she is referred to in her father's will.[8] Samuel and Sally had six known children:

1] Joseph Palmer Orwick was born in November 1819 and he died on 4 January 1850. He was married first to Rachel Blackledge on 21 May 1840[9] in Columbiana County. After she died, he married Elizabeth Dunlary on 17 April 1843[10] in Columbiana County.

2] Mary Orwick was born in August 1823 and she married John H. Brown on 12 February 1846[11] in Jefferson County.

3] Henry Clay Orwick was born on 17 January 1829 and he died on 16 April 1923.[12] He was married to Emily Ann Leonard on 6 February 1850[13] in Jefferson County and then to Margaret Jane Maple on 8 January 1857[14] in Carroll County.

---

[5] *Register of Enlistments in the U.S. Army, 1798-1914*; Philip Carroll, 2nd Regiment of U.S. Artillery, National Archives Microfilm Publication M233; Records of the Adjutant General's Office, 1780's-1917, Record Group 94; National Archives, Washington, D.C.

[6] *Pension List of 1835*, page 163, Ohio, Jefferson County, Private Philip Carroll, 2nd Regiment of Artillery.

[7] Marriage of Samuel Orwick and Sarah Parmer, 11 November 1817, volume 1, page 312, by Joseph McLaughlin, Justice of the Peace, Columbiana County Probate Court, Lisbon, Ohio.

[8] Will of Joseph Palmer, written 27 September 1847, will book, pp. 438-442, Carroll County Probate Court, Carrollton, Ohio.

[9] Marriage of Joseph P. Orwick and Rachel Blackledge, 21 May 1840, volume 1, page 147, return 955, by David Watt, Justice of the Peace for Fox Township, Columbiana County Probate Court, Lisbon, Ohio.

[10] Marriage of Joseph Orwick and Elizabeth Dunlavy, 17 April 1843, volume 3, page 338, by Thomas G. Huston, Justice of the Peace, Columbiana County Probate Court, Lisbon, Ohio.

[11] Marriage of Mary Orwick and John Brown, 12 February 1846, book 6, page 114, return 341, by C. E. Weirich, Minister of the Gospel, Samuel Orwick sworn, Jefferson County Probate Court, Steubenville, Ohio.

[12] Death certificate of Henry Clay Orwick, 16 April 1923, Fox Township, Carroll County, Ohio, file number 24034, Ohio Bureau of Vital Statistics, Columbus, Ohio.

[13] Marriage of Henry Orwick and Emily Ann Leonard, 6 February 1850, book 6, page 389, return 131, by John Allmon, Justice of the Peace, Jefferson County Probate Court, Steubenville, Ohio.

4] William T. Orwick was born in 1830 and he died in 1860. He married Sarah Deets on 24 April 1856[15] in Carroll County.

5] Sarah Jane Orwick was born around 1831 and she married Hugh Divin on 18 July 1871[16] in Tuscarawas County, Ohio. She died in 1872.

6] Margaret Orwick was born around 1839. She married Joseph Miller on 17 June 1858 in Jefferson County.

The family of Samuel Orwick was living in Knox Township, Jefferson County, during the 1820 census.[17] There is one child under 10 years of age in this return who would have been Samuel's son Joseph. Mary and Margaret Carrell have not been found in this census.

By 1825 the family had relocated to Clinton Township, Jefferson County, which had been formed from part of Knox Township. The personal property tax for 1826 lists Margaret Curry and Samuel Carrel.[18] This Samuel Carrel is Samuel Orwick since his mother had no other sons named Samuel. Margaret had married George Curry on 12 February 1822 and he had died by 1826.[19] They had two children: Isaac and Rebecca. Sally Palmer's (wife of Samuel) sister, Lavina Palmer, married Joseph Carrel, the brother of Margaret Carrel and half-brother of Samuel, on 3 June 1827 in Columbiana County, Ohio.[20]

The 1827 personal property tax for this Clinton Township only assessed Samuel Orwick.[21] The 1828 personal property tax list once again lists Margaret Curry and Samuel Carrel (sic).[22] Margaret is listed in the 1829 personal property tax list but not Samuel.[23]

---

[14] Marriage of Henry C. Orwig and Margaret Jane Maple, 8 January 1857, volume II, page 224, Carroll County Probate Court, Carrollton, Ohio.

[15] Marriage of William Orwick and Sarah Deets, 24 April 1856, volume II, page 205, Carroll County Probate Court, Carrollton, Ohio.

[16] Marriage of Hugh Divine and Sarah Orwick, 18 July 1871, volume 6, page 495, license number 12172 dated 10 July 1871, married by James Truman, justice of the peace for Warren Township, Tuscarawas County Probate County, New Philadelphia, Ohio.

[17] *1820 Federal Census*, Ohio, Jefferson County, Knox Township, page 270, Samuel Orwick; National Archives Microfilm M-33, Roll 91.

[18] Clinton Township Number 5, taxes on personal property for the year 1826, Margaret Curry and Samuel Carrel, Auditor's Office, Jefferson County, Steubenville, Ohio.

[19] Marriage record of George Curry and Margaret Carrel, marriage book 2 (1813-1824), page 254, Jefferson County Probate Court, Steubenville, Ohio.

[20] Marriage record of Joseph Carrel and Lavina Palmer, marriage book 2 (1818-1833), page 271, Columbiana County Probate Court, Lisbon, Ohio.

[21] Clinton Township Number 5, taxes on personal property for the year 1827, Samuel Orwick, Auditor's Office, Jefferson County, Steubenville, Ohio.

[22] Clinton Township Number 5, taxes on personal property for the year 1828, Margaret Curry and Samuel Carrel, Auditor's Office, Jefferson County, Steubenville, Ohio.

[23] Clinton Township Number 5, taxes on personal property for the year 1829, Margaret Curry, Auditor's Office, Jefferson County, Steubenville, Ohio.

The Orwick family is listed in the 1830 census returns of Clinton Township.[24] The children who were added to the family since 1820 were Mary, Henry and William. Besides Samuel is living his mother, Mary, and his half-sister, Margaret Curry. Mary and Margaret are listed in both the 1830 and 1832 tax lists but only Margaret is listed in the 1833 list.[25] Mary may have died in 1832.

Clinton Township is a defunct township which originally occupied township 12 of range 4 in the Seven Ranges of Ohio. Today, this former township is part of Lee and Loudon Townships in Carroll County and Springfield Township in Jefferson County.

The *Carroll County, Ohio 1833 Tax List* [26] shows that Samuel owned a town lot in Woodberry, Jefferson Township, Carroll County, and land in Washington Township, Carroll County. Woodberry was plated in 1832 but it never developed into a community. No land deeds have been found for Samuel Orwick at the Recorder's Office's in either Carroll County or in Columbiana County. The Carroll County tax list shows that Samuel owned land in section 8 of township 14 in range 5 and that he had two horses and three head of cattle.

The 1833 property tax list for Jefferson Township shows that Samuel actually paid taxes on four (4) lots in Woodbury.[27] Samuel paid personal property taxes in Washington Township, Carroll County, in the same year.[28] This is where the family was living in 1832. Samuel continued to pay taxes on his properties between 1834 and 1838.[29]

The family had moved to Fox Township, Carroll County, by the time of the 1840 census.[30] Added to the family since the last census were three girls and a boy. Two of the girls were probably Sarah Jane and her sister Margaret. The identity of the other girl and the boy is unknown, possibly, two other children or grand children.

---

[24] *1830 Federal Census*, Ohio, Jefferson County, Clinton Township, page 61, Sam'l Orwick, Mary Carrel and Margaret Curry; National Archives Microfilm M-19, Roll 134.

[25] Clinton Township Number 5, taxes on personal property for the year 1830, Mary Carrel and Margaret Curry, Auditor's Office, Jefferson County, Steubenville, Ohio.

Clinton Township Number 5, taxes on personal property for the year 1832, Mary Carrel and Margaret Curry, Auditor's Office, Jefferson County, Steubenville, Ohio.

Clinton Township Number 5, taxes on personal property for the year 1833, Margaret Curry, Auditor's Office, Jefferson County, Steubenville, Ohio.

[26] Bell, Carol Willsey, C.G., *Carroll County, Ohio 1833 Tax List*, (Columbus, Ohio, 1980), page 3, Jefferson Township, Samuel Orwick, owned lot in Woodburg or Woodbury, page 17, Washington Township, Samuel Orwick, range 5, township 14, section 8, 2 horses, 3 cattle.

[27] Jefferson Township, taxes on property for the year 1833, Samuel Orwick, Auditor's Office, Carroll County, Carrollton, Ohio.

[28] Washington Township, taxes on personal property for the year 1833, Samuel Orwick, Auditor's Office, Carroll County, Carrollton, Ohio.

[29] Washington Township, taxes on personal property for the year 1834, Samuel Orwick, Auditor's Office, Carroll County, Carrollton, Ohio.

Jefferson Township, taxes on personal property for the year 1835, Samuel Orwick, Auditor's Office, Carroll County, Carrollton, Ohio.

Lee Township, taxes on personal property for the year 1836, Samuel Orwick, Auditor's Office, Carroll County, Carrollton, Ohio.

Lee Township, taxes on personal property for the year 1838, Samuel Orwick, Auditor's Office, Carroll County, Carrollton, Ohio.

[30] *1840 Federal Census*, Ohio, Carroll County, Fox Township, page 218, line 18, Samuel Orwick; National Archives Microfilm M-704, Roll 381.

The family was living in Brush Creek Township, Jefferson County, in 1850.[31] Samuel's wife, Sarah, had died before the 1850 census was recorded so she is not listed in this return. Listed with Samuel are his children Henry, Sarah, William and Margaret. Samuel is listed as a 54-year old laborer, having been born in Pennsylvania. Henry is listed as a shoemaker. Samuel's grandson, Wesley Orwick, is living with Samuel's half-sister, Margaret Curry, during the 1850 census returns.

Samuel remarried to Rebecca Elliott on 31 October 1851[32] in Fox Township, Carroll County. She was the daughter of Jeremiah Elliott and Catherine Morgan. Samuel and Rebecca had four children.

1] James H. Orwick was born on 7 August 1851 and he died on 10 January 1917. He married Elizabeth R. Thompson on 26 June 1874 in Jefferson County.

2] Silas Orwick was born in November 1852 and he married Sarah Smith on 6 May 1899 in Jefferson County.

3] Amos Orwick was born in August 1856 and he married Martha J. Cantlen on 21 June 1876 in Jefferson County.

4] Matilda Orwick was born around 1862. It appears that she died before 1880.

Rebecca brought into the family two children who appear to have been illegitimate since they both used the Elliott surname and not their father's surname. Samuel's step-children were:

1] Henry B. Elliott was born in August 1844.

2] Rosanne "Annie" Elliott was born in July 1846.

In 1850 Rebecca was living with her mother, Catherine, her brother David, and her two children in Fox Township.[33] Henry and Annie are listed as Elliotts.

The Orwick family was living in Salem Township, Jefferson County, during the 1860[34] census. Their post office is listed as Richmond, Ohio. With Samuel and Rebecca are Henry, Ann, James, Amos and Silas. Henry and Ann are listed as Orwicks and not Elliotts. The family was living in Island Creek Township, Jefferson County, during the 1870[35] returns. Samuel is listed as a farmer with personal property worth $350. With Samuel and Rebecca are their children James, Amos, Silas and Matilda.

No death record has been found for Samuel as well as his cemetery location. Also, no will or administration letters have been found. Rebecca is living with her son, Amos, and his family in Salem

---

[31] *1850 Federal Census*, Ohio, Jefferson County, Brush Creek Township, 6 September 1850, page 463, dwelling 1192, family 1333, lines 25-26, Samuel Orwick; National Archives Microfilm M-432, Roll 699.

[32] Marriage of Samuel Orwick and Rebecca Elliott, 31 October 1851, volume II, page 62, Carroll County Probate Court, Carrollton, Ohio.

[33] *1850 Federal Census*, Ohio, Carroll County, Fox Township, 5 October 1850, page 172B, dwelling 2356, family 2361, lines 12-16, Rebecca Elliott; National Archives Microfilm M-432, Roll 664.

[34] *1860 Federal Census*, Ohio, Jefferson County, Salem Township, 20 July 1860, page 285, dwelling 838, family 829, lines 25-31, Sam'l Orrick; National Archives Microfilm M-653, Roll 993.

[35] *1870 Federal Census*, Ohio, Jefferson County, Island Creek Township, P.O. Salineville, 1 June 1870, page 322, dwelling 175, page 162, Samuel Orwick; National Archives Microfilm M-593, Roll 1228.

Township, Jefferson County, during the 1880[36] census. Rebecca died before 1900.

Samuel filed for a military pension on 18 March 1871 stating that he had volunteered to serve in the Ohio militia during the War of 1812.[37] He wrote that he enlisted under Lieutenant Rieley and that he served in Captain Stockton's company. On 26 July 1871, the U.S. Treasury Department declared that Orwick was not listed on the muster roles of Captain William Stockton's company of the Ohio militia and his pension claim was rejected.

He re-filed for a pension on 25 March 1871 (before the government rejected his first application) with his correct military service information. Once the government saw his new petition, his pension application was approved and he was admitted to the pension roles on 24 November 1871. He collected eight dollars per month.

Samuel died on 5 March 1873 in Richmond, Salem Township, Jefferson County. His name was dropped from the pension rolls on 27 March 1873. After the death of her husband, Rebecca did not apply for a widow's pension, for which she was entitled. She was declared an imbecile on 21 February 1881 by the Jefferson County Probate Court and J. W. Kirk was appointed as her guardian.

With Rebecca having no source of income, Kirk on her behalf applied for a widow's pension which was issued to her on 13 April 1881. The pension was made retroactive from 9 March 1878. She received eight dollars per month. In the widow's pension it is stated that Miss Parmer (sic) was Samuel's first wife.

A death notice for Rebecca Orwick was printed in *The Steubenville Herald* of Steubenville, Ohio, on Tuesday, 23 February 1897.[38] Rebecca had died the previous day in her home near East Springfield in Salem Township. The notice lists her children as Henry, Annie, James and Silas.

# Ohio governors who served in the War of 1812

Eight Ohio governors participated in the War of 1812 either in the Ohio militia or in the U.S. Army. Samuel Huntington, who was the third governor between 1808 and 1810, was the 8th Military District Paymaster for the United States Army. Fort Huntington in Cleveland was named in his honor.

Return Jonathan Meigs, Junior, was the fourth governor of Ohio from 1810 to 1814 and as such, he was the commander of the Ohio militia during most of the war. He received an appointment as the Postmaster General of the United States in 1814 so the fifth governor, Othniel Looker, assumed the duties of the militia commander for the rest of the war. Fort Meigs was named in honor of Governor Meigs.

Allen Trimble was the governor of Ohio twice. He served as the eighth governor in 1822 and as the tenth governor between 1826 and 1830. He was a colonel in the Ohio militia during the war.

The eleventh governor was Duncan McArthur who served between 1830 and 1832. He was a major general in the Ohio militia and he was commissioned a brigadier general in the U.S. Army. He became the commander of the Army of the Northwest after Major General William Henry Harrison resigned his commission. Fort McArthur was named for Duncan McArthur.

Robert Lucas was a brigadier general in the Ohio militia and he was elected as the twelfth governor of Ohio. He served from 1832 until 1836. Joseph Vance was a captain in the militia and he served as the thirteenth governor between 1836 and 1838. The eighteenth governor, Mordecai Bartley, was the governor from 1844 until 1846. He was a captain and an adjutant in the Ohio militia.

---

[36] *1880 Federal Census*, Ohio, Jefferson County, Salem Township, sheet 19, line 38, Amos Orwick; National Archives Microfilm T-9, Roll 1037.

[37] Samuel Orwick's Pension Application, number SO-4092, SC-8800, WO-40588, WC-31578, 24 November 1871, War of 1812, War Department Collection, National Archives & Records Administration, Washington, D.C.

[38] *The Steubenville Herald*, Tuesday, 23 February 1897, page 4, column 3, Mrs. Rebecca Orwick, Steubenville, Ohio

## Captain Thomas Ramsey of Cincinnati

Captain Thomas Ramsey is one of the more distinguished officers from Ohio who served in the U.S. Army during the War of 1812. He was stationed before and during most of the war in Cincinnati, Ohio, serving as a recruiting officer, but when he was called to duty, he served in Illinois, in Upper Canada (Ontario), and in New York State.

He was born circa 1777 in York County, Pennsylvania, raised in Kentucky and Ohio, married Elizabeth Crane, and settled in Cincinnati prior to 1806. Thomas was commissioned as a second lieutenant on 27 January 1809 in the Regiment of U.S. Rifles. He served as a recruiter from 1809 to 1812 in both Cincinnati and across the Ohio River from Cincinnati at the Newport army barracks. He was promoted to first lieutenant on 31 July 1810 and then to captain on 30 November 1812.

In the spring of 1812, Captain Ramsey was ordered to take his recruits to Fort Russell, Illinois Territory, to man this fort. The fort served as a rendezvous for Illinois militia regiments and as the frontier headquarters for the territorial governor. It was located where Edwardsville now stands.

Captain Ramsey was then ordered to take his company to Portage de Sioux, Missouri Territory, in June 1813 by Brigadier General William Howard, the U.S. Army general in charge of the defenses in the upper Mississippi Valley. General Howard transferred Ramsey's enlisted men to three companies of the 1st Regiment of U.S. Infantry, commanded by Captains John Symmes, Simon Owens, and Horatio Stark. The general then ordered Ramsey and his officers back to Cincinnati to recruit more men.

Ramsey returned to Cincinnati and he started the task of recruiting another rifle company. He did not return to Missouri until after the war. By the end of the summer Ramsey had recruited his second company, and had joined Colonel Thomas Smith of the Rifle Regiment on their march towards northern Ohio to participate in Major General William H. Harrison's invasion of Upper Canada.

Colonel Smith's detachment of riflemen from Ohio, Kentucky and Tennessee numbered 520 men. They arrived too late to join Harrison's army when the Army of the Northwest invaded Upper Canada on 27 September 1813. The riflemen, however, were used as occupation troops at Amherstburg, Upper Canada, which is on the lower Canadian side of the Detroit River.

In late 1813, General Harrison was ordered by the Secretary of War to take half of his men to New York State in order to support the Army of the North. Colonel Smith distributed the men in Captain Ramsey's company to the other companies under his command, and then ordered Ramsey back to Cincinnati to recruit another company.

In the summer of 1814, Captain Ramsey received his third set of marching orders to take his third company to New York State. Upon reaching Fort Stephenson (now Fremont, Ohio), the company was transported by U.S. naval ships to Fort Erie, Upper Canada, across the Niagara River from Buffalo, New York, arriving on 30 August 1814. On 17 September 1814 the company participated in the American attack on British positions which were besieging the fort. Captain Ramsey was wounded during this action which is called the Battle of Fort Erie.

Thomas Ramsey was discharged from the army on 15 June 1815 when the wartime army was reduced to its peacetime strength. He was recalled to military duty on 2 December 1815 and ordered to report for duty to Fort Belle Fountaine, outside of St. Louis.

Captain Thomas Ramsey met an untimely death in 1818 at the hands of Captain Wyly Martin, also of the Rifle Regiment, on Bloody Island off St. Louis in the Mississippi River. Ramsey was shot in a duel with Martin on 6 August 1818 and he died a few days later.

Ramsey was given a full military and Masonic funeral on 17 August 1818. He was a registered member of Cincinnati Lodge No. 13 and was a visitor at Western Star Lodge No. 107 at Kaskaskia, Illinois. The Missouri Lodge No. 12 of St. Louis performed the services. It is assumed that he was buried at Fort Belle Fountaine.

Thomas' family remained in Cincinnati throughout the War of 1812 and may have accompanied him to St. Louis after the war. Elizabeth never remarried. No military pensions or land bounties have been found for Thomas or for his widow. There was a 'Elizabeth Ramsey' who was enumerated in Cincinnati during the 1830 census. There was also a John Ramsey living beside her who is probably a son.

Elizabeth Ramsey died in 1849 and she was buried in Spring Grove Cemetery in Cincinnati. She was buried in the same cemetery lot as her daughter, Eliza, and son-in-law William Gockel. William and Eliza had a son named Thomas Ramsey Gockel.

# The Incident at Goshen

The Indian mission at Goshen was the subject of many verbal attacks and threats throughout the War of 1812. The mission was the home to a small band of Christian Indians who were trying to remain neutral during the war. The Moravian Church under the missionary Rev. Benjamin Mortimer operated the mission. In August 1812 there were seven Indian men and their families living in Goshen.

Rev. Mortimer feared for the lives of his Indians. He received many threats that the local white inhabitants where were going to kill the Indians. Many of the threats came from New Philadelphia, which lay across the Tuscarawas River and three miles northeast of the mission.

The Indians in the northwest area of the country were on the rampage again killing white settlers. Brigadier General William Henry Harrison had subdued the Indians the previous year at the Battle of Tippecanoe but with the start of the war the Indian war parties were on the move again.

The citizens of Tuscarawas County and particularly New Philadelphia feared that the Indians of Goshen and their friends from the west were going to massacre them. The Goshen Indians were from the Delaware tribe who had family and friends in the Indian territory beyond the Greenville Treaty Line. These other Indians sided with the British during the war.

Warner's *History of Tuscarawas County, Ohio,*[39] does tell the story of the incident at Goshen but the story is very misleading. The timeline of Warner's story states that "shortly after the surrender of Hull, three Indians, said to be unfriendly, arrived at Goshen". The story gives the impression that the incident was over and done with shortly after August 16th when Hull surrendered his army.

In actuality the incident started in July 1812 and continued to 2 April 1813 when four Indians were arrested. Months later under armed guard, three of the Indians would be transported and then released in western Ohio. The incident at Goshen played out for almost a year.

## Rev. Mortimer's Letters

### First Letter – 1 August 1812

Rev. Benjamin Mortimer wrote three letters to Governor Meigs seeking protection from the state for his Indians. These letters, dated 1 August, 8 August and 21 August 1812, contain a wealth of information on the mood of the county and on the local militia.

As stated, Mortimer feared for the lives of his Indians and he did not trust the citizens of the county, particularly the local militia. The main purpose of Mortimer's letters was to have the governor order another militia company from another part of the state to this county in order to protect the mission from the local population.

He wrote that the military draft was already in the making and he feared for the mission's safety once the militia left the county. Mortimer states in his letter that the militia was in a state of uneasiness before August 1st. With war fever beginning to take effect in all parts of the country the men had been getting ready for war and for their eventual call up for military duty.

---

[39] *The History of Tuscarawas County, Ohio,* (Warner, Beers, and Company, Chicago, Illinois: 1884), chapter XII, Tuscarawas County in the Wars of 1812 and 1846, pp. 407-409.

The missionary's biggest concern was that after the militia left the county then the mission would be unprotected from the local population. He also feared that after the war had started he would have many Indians from the northwest who would be seeking asylum at his mission. It is assumed that Rev. Mortimer feared that another Gnadenhutten type massacre could happen at Goshen.

### Second Letter – 8 August 1812

Rev. Mortimer begins his second letter by stating that the militia muster had occurred on the preceding Thursday, which places the date as 30 July 1812. He continued by saying that "different militia companies were drafted". At least two volunteer militia companies were organized on this day.

Captain David Casebeer's company was activated on August 1st two days after this muster. Captain George Richardson's Company would be activated on August 29th while Captain Joseph Johnson's Company would begin their military service on October 18th.

Two letters from the citizens of the county to Governor Meigs in September stated that two companies were on active duty in Mansfield while the other was in Urbana. According to Mortimer, as of August 8th the first two companies were still in the county waiting for their marching orders.

The next section of the letter gives an account of an elderly Indian from Greentown, near Mansfield, who was last seen on July 4th wearing his red coat and carrying a British rifle. The Indian was traveling down the Tuscarawas River and heading towards the Stillwater Creek to see a blacksmith. He was also boasting to the local citizens of the number of whites that he had killed during the revolution.

The men in the local militia wanted to shoot him as a British spy. They also wanted to kill every Indians in the county to protect their families from harm before they received their marching orders.

Rev. Mortimer stated that Colonel Robert Bay was in New Philadelphia. Bay had talked to Mortimer and said that he would give his support in protecting the Indians at the mission.

### Third Letter – 21 August 1812

On the night of August 8th, apparently after Rev. Mortimer had wrote and sent his second letter to the governor, a Mr. Laffer and a Mr. McConnel came to the mission around midnight to warn Mortimer that 20 armed men were coming to search for Indians.

A report had reached New Philadelphia the day before that fifteen armed Indians were heading for Goshen. Rev. Mortimer, Laffer and McConnel protected the two Indian families that were in Goshen that night while the twenty-armed men searched for Indians in the other buildings in town.

On August 14th the elderly Indian showed up again in New Philadelphia during the mustering of the militia. He showed himself in public and nothing happened to him. Rev. Mortimer states the mode of the people was a lot better.

Mr. Laffer was probably Henry Laffer who was the county sheriff between 1810 and 1813.[40] After the war he moved to Sandy Township where he founded Sandyville. Mr. McConnell was probably Alexander McConnell of New Philadelphia who operated a tailor shop and who was the commander of the local militia cavalry company.[41] This cavalry company was not called to military duty during war but was probably used by the local authorizes to patrol the area of western Tuscarawas County looking for signs of Indians.

## The Citizens ask for Protection

### The First Letter – 5 September 1812

[40] *Ibid*, pp. 365, 468.

[41] *Ibid*, page 469.

By September 5<sup>th</sup> the three volunteer militia companies formed within the county had received their marching orders and had left the area. The citizens of New Philadelphia wrote the first of two letters to Governor Meigs asking for protection from the Indians. In the first letter the citizens stated that they no longer feared the Indians living at Goshen but they feared that these Indians were harboring Indians who were loyal to the British.

One of the volunteer militiaman who had come down the river from Cleveland and who was with General Hull's army stated that he recognized one of the Indians at Goshen from the Battle of Malden. The petitioners requested protection from the governor since the three companies had left the county with nearly all of the able-bodied men and most of the weapons and powder.

## The Second Letter – 24 September 1812

The county citizens wrote their second letter to the governor after the area went into a crisis when the enemy Indians attacked and killed white settlers near Mansfield. These attacks were reported by Colonel Bay to Governor Meigs on September 12<sup>th</sup>[42]. The first attack, which occurred on Thursday, September 10<sup>th,</sup> killed four people, two men and two women. Two days later on the 12<sup>th</sup> the Indians attacked again and killed six militiamen. They also killed a man and wounded his daughter.

The petitioners stated that nearly all of the settlers had evacuated the western area of the county and they are now in New Philadelphia. They reminded the governor that the county had sent three companies to the west and that they were still without able-bodied men, arms and powder.

*One note:* Colonel Bay wrote his letter from Mansfield and the petitioners stated that he left New Philadelphia on the 24<sup>th</sup>. It appears that Colonel Bay was splitting his time between his troops in Mansfield and trying to keep the peace in New Philadelphia. Apparently Colonel Bay was under orders from the governor to watch Tuscarawas County and to keep him informed on exactly what was happening. As a result, Colonel Bay traveled back and forth between Mansfield where the 4<sup>th</sup> Brigade was operating and New Philadelphia.

## A Change in Tactics

The citizens of the county were not happy that the governor had ignored their pleas for protection so they decided to send representatives to Lieutenant Colonel Isaac Van Horne, the state adjutant general, at Zanesville.

On 1 October 1812 Van Horne wrote to Governor Meigs in his report on the happenings within the 4<sup>th</sup> Brigade.[43] He told the governor that on September 30<sup>th</sup> George Dickey and Henry Sherman from Tuscarawas County along with 12 other citizens had met with him and they had asked for the release of some of the county's militiamen who were stationed at Mansfield.

Two reasons were given. The first reason was that citizens were starting to see Indian trails five to seven miles from New Philadelphia and that some of the citizens are starting to flee to New Philadelphia again. The second reason was that the county still needed protection and that the men were needed to harvest the crops. In Colonel Van Horne's postscript he said that a number of families from the Sugar Creek area had left for fear of the Indians.

[42] *Return Jonathan Meigs, Jr., and the War of 1812, Volume II of the Document Transcriptions of the War of 1812 in the Northwest,* transcribed by Richard C. Knopf, (The Anthony Wayne Parkway Board, The Ohio State Museum, Columbus, Ohio, 1961), page. 50, Robert Baylor to Meigs, 12 September 1812.

[43] *Ibid*, page 150, Van Horne to Meigs, 1 October 1812.

On October 29th Captain George Richardson's Company was released from duty followed two days later by the release of Captain David Casebeer's Company. A good portion of the Tuscarawas County militia was back home where the situation would remain stable until after the new year.

## Four Indians are arrested

With the return of two of the militia companies from Mansfield the county remained quite until late winter when four Indians were arrested at Goshen on 2 April 1813. Colonel Bay wrote to Governor Meigs on April 3rd informing him of the arrests and asking him to make a decision as to what to do with the Indians.

Two of the Indians were men from Goshen and they had been harboring the other two men. One of the local men was in irons for safekeeping. The other two Indians had come from Malden, Upper Canada, in order to convince the Goshen Indians to leave the United States and to come to Canada.

One of the Indians was named Philip Connotchy who had been living at Goshen before the war and had not been seen since the war started. The men had British rifles, powder and shot plus hatches, knifes, and war clubs on them. Connotchy admitted that he was at the Battle of Brownstown and at the surrender of Hull. The other Indian said he was at the defeat of General Winchester. Connotchy also admitted that he was with a party of 35 Indians in the fall that attacked Greentown. He said that he was not involved in the murders committed there.

In Warner's *History of Tuscarawas, County, Ohio,* it is told that Captain Alexander McConnell and six of his cavalrymen captured three of the Indians on a small island near Goshen in the Tuscarawas River. The book does not mention a fourth Indian. The captain and his troops escorted the men to the country jail.

As soon as the news of the arrests reached Wooster, forty armed men under Captain Mullen left Wooster for New Philadelphia with the intent of murdering the Indians. Henry Laffer and Alexander McConnell once again came to the rescue and protected these Indians.

Later, Governor Meigs arrived at New Philadelphia and he instructed Lieutenant Shane, the recruiting officer, to take the Indians to Zanesville. From there the Indians were taken to General Harrison's headquarters at Seneca where the Indians were released. Lt. Shane was Abraham Shane of New Philadelphia.

# The Mystery of Abraham Tope

They are certain genealogical information which are found in family histories that seems to perpetuate themselves over the generations. What one person researched and placed into print years ago never seems to change. Later generations copies the information, cites the sources and adds no further information in order to enhance the original research.

The mystery as to what happened to Abraham Tope of Jefferson County, Ohio, has persisted for over a hundred years, maybe longer. According to Melancthon Tope's *History of the Tope Family*,[44] first published in 1896 and later revised by Austin Dale Maddux in 1980, Abraham was the son of John Tope (1767-1844) and Mary Helmick, no birth or death dates.

In his book, Tope simply says, "Abram was the oldest son. He went to the war of 1812 and died in the war – thought to have been killed in battle. Some think he was a single man; by others it is claimed he had a wife and two children, but as to what has become of them deponents say not." This is the passage which is repeated in later family histories. A family group sheet in Tope's book lists Abraham as "Abram" and having died in 1812.

Later histories claim that Abraham had married Elizabeth Roop on 30 April 1811 in Jefferson County.[45] Still others claim that Levi Tope was one of the two children of Abraham.

This article will determine when Abraham died and who were his wife and children. Furthermore, it will also try and ascertain if he served during the War of 1812, if he died during this war, and what happened to his family after his death.

According to the Jefferson County Probate Court, Abraham had no will recorded. The Jefferson County Court of Common Pleas does, however, have a Petition for Partition for Abraham Tope's property.[46] During the April 1816 Term, Jacob Roop and Elizabeth Tope, widow of Abraham Tope, petition the court to divide the property of Abraham Tope. Abraham had owned 278 acres of land in the northern half of section 27 in township 10 of range 3 of the Steubenville Land District. This property is in today's Salem Township in Jefferson County.

The court appointed Henry Meese, Jacob Lease and Thomas Cole to study the property in order to determine if it could be sub-divided. The men returned their findings to the court and they stated "that the property can not be divided without injury or prejudice to the whole." They valued the property at $6.50 per acre.

This document proves that an Elizabeth was indeed the wife of Abraham Tope and that Abraham had died before April 1816. Jacob Roop was probably the court appointed executor of Abraham's estate. Jacob appears to be either the father or a brother of Elizabeth Roop.

For his military service, the *Roster of Ohio Soldiers in the War of 1812* shows that there was an "Abraham Toppe" who served as a private in Captain David Peck's Company from Jefferson County.[47] The company was activated on 25 August 1812 and it was discharged on 28 February 1813. The company was attached to Lieutenant Colonel John Andrew's 1st Regiment in the 2nd Ohio Brigade.

This brigade was used as a construction unit building a military road from Canton, Ohio, thorough Wooster and Mansfield, to Fort Ferree in what is now Upper Sandusky, Ohio. From Fort Ferree, the road was extended to Fort Meigs, just south of Toledo, Ohio.

---

[44] Tope, Melancthon, *History of the Tope Family*, (The Patriot Office: Bowerston, OH, 1896), revised in 1980 by Austin Dale Maddux, pp. 22 and 25.

[45] Marriage of Abraham Tope and Elizabeth Roop, 30 April 1811, marriage book 1, number 186, Jefferson County Probate Court, Steubenville, Ohio.

[46] Petition for Partition, Jacob Roop and Elizabeth Tope, widow of Abraham Tope, Common Pleas Journal B, April Term 1816, pp. 249, 271, Jefferson County Court of Common Pleas, Steubenville, Ohio.

[47] *Roster of Ohio Soldiers in the War of 1812*, Adjutant General of Ohio, 1916, (Clearfield Co., Baltimore, Maryland: 1968), page 9, Capt. David Peck's Company, Jefferson County, 25 Aug 1812 - 28 Feb 1813, Private Abraham Toppe.

Abraham's father, John Tope, and his uncle, George Tope, settled in Jefferson County before 1806. George had no sons by the name of Abraham according to the *History of the Tope Family* so this "Abraham Toppe" was probably the Abraham of this article.

Obtaining the military service records of Abraham Tope from the National Archives for the War of 1812 does show that his surname was spelled as "Tope" so the entry in the *Roster of Ohio Soldiers in the War of 1812* is in error.

In his military service records, there is no mention of Abram's death while on military duty. On the Company Muster Roll, dated for the period of 25 August to 30 November 1812, it shows that Abraham was called into the service on 21 September 1812 and that he was discharged on 21 March 1813. He was present for duty on 21 March 1813.[48]

On the Company Pay Roll report, dated for the period 25 August to 30 November 1812, it shows that Abraham served for two months and ten days between 21 September and 30 November 1812. He was paid $15.54 and he had not picked up his pay.[49] This failure to pick up his pay does not prove that he died during his tour of duty.

What is missing in his military service records are a company muster roll for his service between November 1812 and March 1813 and another company pay roll report for the same time period. Since the first report shows that he was 'present' on 21 March 1813 at the end of this tour of duty, it appears that Abraham died after he had returned home from military duty.

Abraham is not listed on the Pension Roll of 1818 as having been killed or dieing in the war. If he had died on duty then his wife and children would have qualified for a pension. No pension has been found. The National Archives also has no record of him serving in the U.S. Army after his tour of duty ended with the militia.

If Abraham had died on military duty, either with the militia or the U.S. Army, his children would have been required to have a court appointed guardian in order for his wife and children to have qualified for a federal pension. There are no guardianship papers for the heirs of Abraham Tope at the Jefferson County Courthouse. If there were papers, these documents would have shown when and now he died, who the guardian was, and they would have contained the names and ages of his children.

Abraham Tope died between 21 March 1813 and the April Term of the Common Pleas Court in 1816. He probably died with his family in Jefferson County and he may have been buried on his farm. There is no listing for him in the cemetery books from Jefferson County.

John Tope, the father of Abraham, died on 26 October 1844 in Salem Township, Jefferson County, Ohio. He was buried in the Lease's Cemetery in Salem Township. His will was written on 8 March 1844 and it was probated on 18 November 1844.[50]

Only ten of John's fourteen children are mentioned in his will. Of the four missing children: Abraham had died before 1816; Henry had moved to Illinois and he had no heirs: Elizabeth, who married Enoch Hough, had died in 1835; and an unnamed male child had died very young.

Two grandchildren are mentioned in his will: Levi Tope and Susanna Houzer (probably Houser). They can not be the heirs of Elizabeth since the children's surnames are not Hough. They can not be the children of Henry, since he is supposed to have been childless. This leaves Levi and Susanna as the children of Abraham and Elizabeth Tope. Finally, it appears that Susanna was married so she probably was not a daughter of one of John's younger female children.

---

[48] Company muster roll of Captain David Peck's Company of Ohio Militia, for 25 August 1812 to 30 November 1812, Abram Tope, card number 37728292, roll box 210, roll record 1245, War of 1812, Record Group 94, National Archives and Records Administration, Washington, D.C.

[49] Company pay roll of Captain David Peck's Company of Ohio Militia, for 25 August 1812 to 30 November 1812, Abram Tope, card number 37728360, roll box 210, roll record 1245, War of 1812, Record Group 94, National Archives and Records Administration, Washington, D.C.

[50] Will of John Tope, will book IV, pp. 63-65, Jefferson County Probate Court, Steubenville, Ohio.

By law, Levi and Susanna could not have been listed in this will if their father or mother, a child of John Tope, was still living. They could have been listed if they were being given certain items promised to them by their grandfather. There is no indication of this in the will. Levi and Susanna received their father's inheritance.

Elizabeth, along with her two children, moved to German Township, Coshocton County, Ohio, probably with her brother or father, Jacob Roop. This is where she married John Farver on 3 November 1816.[51] Both the Roop and the Farver families are listed in the 1820 census for German Township.

In this census John and Elizabeth are listed in the 25 to 45 year range while there is one son listed in the under 10 age range while three daughters are listed in the under 10 age range for females. At least two of these children were probably the children of Abraham Tope.

The northern portion of Coshocton County became a part of Holmes County in 1824. German Township was transferred to this new county. The Farver's are listed in the census returns for Holmes County starting in 1830.

John and Elizabeth are buried in the New Bedford Cemetery in Clark Township, Holmes County.[52] John died on 7 March 1862 at the age of 70 while Elizabeth died on 14 April 1878 at the age of 82 years, 7 months and 3 days. She was born on 11 September 1795 (calculated from her tombstone).

Levi Tope was married to Susannah Gonser on 19 September 1830 in Holmes County and it appears that she died before 16 April 1844 when Levi remarried to Mary Ann Correll.[53] It appears that Susanna was married to Jacob Howser (sic) on 10 June 1832 in Holmes County.[54]

John and Elizabeth Farver are still living in German Township in 1850. The federal census returns for this year shows that there are ten children in the household.[55] They have a son named Abraham who was 22 years old. Levi and Mary Tope are living in Monroe Township in 1850.[56] Among their seven children is their 16-year old son Abraham. Jacob and Susanna Houser have not been found in this census.

In conclusion, Abraham Tope did not die while serving with the U.S. military forces during the War of 1812 although he may have died in the later part of the war. He was certainly dead by April 1816. He was married to Elizabeth Roop and they had two children, Levi and Susanna Tope. It is hoped that in the future other documents will surface which will list his death date and cause of death.

---

[51] Hunter, Miriam C., *Marriage Coshocton County, Ohio, 1811-1930*, volume 1, (Coshocton, OH: Coshocton County Public Library, 1967), marriage of John Farver and Elizabeth Tope, 3 November 1816.

[52] Dickinson, Marguerite & Richard H., *Holmes County Cemetery Records*, (Millersburg, OH: 1970), burial listing for John and Elizabeth Farver, page 111.

[53] Raber, Nellie M., *Holmes County, Ohio, Marriages*, (1954), marriage of Levi Tope and Susannah Gonsor, 19 September 1830, and the marriage of Levi Tope and Mary Ann Correll, 16 April 1844.

[54] *Ibid*, marriage of Susannah (-----) and Jacob Howser, 10 June 1832.

[55] 1850 Federal Census (Free Schedule), Ohio, Holmes County, German Township, John Farver, page 220, dwelling 1892, family 1926, lines 1-12; National Archives Microfilm M-432, Roll 696.

[56] *Ibid*, Monroe Township, Levi Tope, page 138, dwelling 777, family 793, lines 20-28; National Archives Microfilm M-432, Roll 696.

# A Case Study:
## Finding the military records of Alexander Mason for the War of 1812

All brick walls in genealogy will eventually crumble down! Whether you happen to find the right person, at the right time, with the right information, or after you have spent hundreds of hours doing your own research. Finding the military records of an ancestor can be trying, especially for the War of 1812. Many veterans simply do not have records, others have records which can be found, but through unusual avenues.

This story begins during the Ohio Genealogical Society's 2008 Conference in Cincinnati, Ohio, at one of the banquets when I had sat down at a table, which already had seated the family of Melissa Danielsson. Melissa was with her husband, Magnus, and children Emily and Eric. After the usual introductions and greetings, Melissa noticed my War of 1812 badge on my blazer, she smiled and said that she had a relative who had died in this war by the name of Alexander Mason. I returned the smile and said that Alexander had been killed on September 29, 1812, during the Battle of Marblehead Peninsula.

As luck would have it, I had been researching this battle. I had recently submitted an article on this battle to the *Journal of War of 1812*[57] in which Alexander had been killed so I knew about him. I also had presented a paper on this battle during the 8th National War of 1812 Symposium[58] in Baltimore, Maryland. Besides, the September date was easy for me to remember since it is my birthday, although I was born a little after 1812.

Melissa wanted to join the *National Society United States Daughters of 1812*[59] but she could not find Alexander Mason's military records. I, myself, am a member of the *General Society of the War of 1812*[60], the male counterpart to the *Daughters*. Melissa asked me to help her find the military records on her ancestor so she could submit an application for membership. Now, being weak at the knees and always ready to do more research work, I accepted the challenge. Melissa's brick wall was about to crumble!

I knew that this task would be difficult because Mason had been listed as a civilian causality after this battle in an official militia after-action report. He had volunteered to be a scout for a composite company of Ohio militiamen under the command of Captain

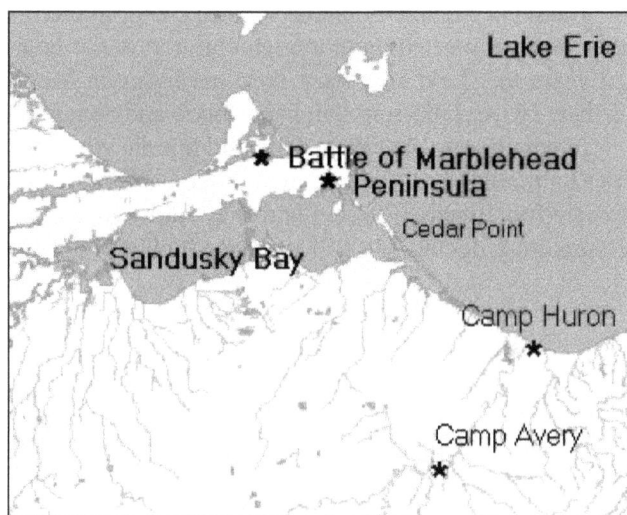

Battle of Marblehead Peninsula, Ohio

Joshua T. Cotton. He was killed during the second skirmish with the Indians.

During my research on this battle, I found that there were a number of civilian scouts who fought in this battle. In this report, the men were called "pilots." Two of these scouts had been killed and two wounded. Questions began to come to mind! Why were these men 'civilians' and not local militiamen? Were they settlers in the area of the battle? Would there be both military records and pension records for these men?

---

[57] *The Journal of the War of 1812*, volume IX, number 4, Winter 2005-2006, (Baltimore, MD), The Battle of Marblehead Peninsula, by Eric E. Johnson, pp. 2-7.

[58] Eighth National War of 1812 Symposium (War of 1812 Consortium), Baltimore, MD, 9 October 2004.

[59] National Society United States Daughters of 1812, http://www.usdaughters1812.org/.

[60] The General Society of the War of 1812, http://www.societyofthewarof1812.org/.

I had previously tried to obtain a listing of all of the men who had participated in this battle. Captain Cotton's detachment was made up of volunteers from five different militia companies plus civilian scouts. A list of Cotton's detachment has not been found. Alexander Mason is not listed on any of the muster rolls for the militia companies stationed at Camp Avery in Huron County, Ohio. He is also not listed in the *Roster of Ohio Soldiers in the War of 1812*.[61]

Checking *Ancestry.com's* compact disk[62] (CD) on the militia rolls from the War of 1812 showed that the federal government did not have the service record for an Alexander Mason who had served from Ohio.

One other source of information that I had used without success in finding Mason's military papers was the Internet. Long ago, I had found a simple search technique, which I had used repeatedly in a multitude of other projects. When you type in Mason's full name in an Internet browser, you will get hundreds, if not thousands, of web pages returned which has Alexander and/or Mason on its pages.

Using a Boolean search, that is, placing double quotes around Mason's name (example: "Alexander Mason") will return pages which only has this phrase on them. This still returned too many pages so I then used a double Boolean search. Since I was looking for Mason in the War of 1812, I also included the War of 1812 in double quotes (example: "Alexander Mason" "War of 1812"). This gave me five web pages, listing 51 websites, which

Monuments to the battle

had Alexander Mason and the War of 1812 on it. This technique makes Internet browsing very manageable.

Nevertheless, these web pages returned the same information that I had already obtained from using traditional researching methods. At this point in my research, Melissa's brick wall was starting to get a little bit higher.

Now I needed to step back and regroup. What information did I have on Alexander Mason? A number of the men from this battle wrote articles after the war telling of their experiences during this battle. All of them mentioned Mason in their articles. I now had first hand accounts of men who probably witnessed the death of Mason.

In Danbury Township on Marblehead Peninsula (now in Ottawa County, Ohio), Alexander Mason's name is listed on a monument in a small park commemorating this battle. The monument lists the name of the six men killed in the battle and two who had been killed on September 15th. Great information, but this still does not prove that he was a militiaman and that he has a service record. It does help prove that he fought in the war and that he had been killed.

The next step was to go back to the Western Reserve Historical Society's library in Cleveland, Ohio, and revisit the military records of the 4th Division of the Ohio Militia during the war. I had heavily used this source of information when I had prepared my original article on this battle for the *Journal of the War of 1812*.

---

[61] *Roster of Ohio Soldiers in the War of 1812*, Adjutant General of Ohio, (1916, reprinted by Heritage Books, Inc., Bowie, Maryland: 1995).

[62] *Ancestry.com*, (Orem, UT: MyFamily.com, 2000), CD-ROM, <u>Military Records: War of 1812 Muster Rolls</u>.

The men who participated in this battle were under the command of Brigadier General Simon Perkins of the 4th Division. The library's collection for this division lists muster rolls, correspondences, and all types of military records from court-martial papers to forage reports. The only document that I found with Mason's name of it was the after-action report from General Perkins to Ohio's Governor Return Jonathan Meigs, Junior, telling him about the battle and listing the causalities.[63] The militiamen who was killed or wounded were listed by their first and last names while the civilian were only listed by their surnames. Without proof from this collection, it appears that the brick wall was going to last a lot longer!

Turning to the military pension records, I was not able to find any pension listings for Mason's widow in the Pension Lists of 1818,[64] 1835 [65] or 1850.[66] However, a pension was issued to Cornelia, wife of Alexander Mason, according to Virgil White's *Index to War of 1812 Pension Files*.[67] Cornelia received a widow's pension, Old War WF-9923, for Alexander's service as a private in Captain Cotton's Ohio militia company. If there is a pension then there should be military service records.

"Old War" meant that the pension was issued for wounds, injuries, or deaths due to military service during the War of 1812. This pension was not a service pension that was issued after 1871 to surviving veterans or their widows. The monthly amount of these service pensions were based upon the number of days of service of a veteran.

Old War pensions gave the widows the half pay per month of their deceased husbands for five years. Technically, these pensions were not "pensions" but simple compensation for the widows until they remarried or until they have gotten back on their feet financially.

The low number "9923" showed that the pension was issued during or shortly after the war so Cornelia's pension should have been in the 1818 pension list. According to the Pension List of 1818, Ohio had 99 men on the pension roll from the Revolutionary War and 194 men from the War of 1812. Cornelia's application for a pension may still have been pending and it may have been in some state of being processed by the government in 1818.

Two other sources for pensions that were worth checking were White's *U.S. Military Pension Applications of Remarried Widows for Service Between 1812 & 1911*[68] and Craig Scott's *The "Lost" Pensions*.[69] I did not expect to find Cornelia's pension in White's second index but I still wanted to check. This listing was for the rejected pension applications of widow's who remarried after the death of their first husbands who had died during the war. These women never applied for a pension when they were required to submit an application. They applied for a pension after their second marriage which automatically disqualified them for a pension.

---

[63] MSS-660 – Collection of Papers, 1792-1876, Container 1, 3 October 1812, Letter from Brigadier General Simon Perkins to Governor Return Jonathan Meigs, Junior, Western Reserve Historical Society, Archive Library, Cleveland, Ohio.

[64] *Pensioners of the U.S., 1818, Invalids & Half Pay Pensioners, Widows & Orphans*, (Washington, D.C.: E. De Krafft, 1818).

[65] Report from the Secretary of War in Obedience to the Resolutions of the Senate of the 5th and 30th of June, 1834, and the 3rd of March, 1835, in Relation to the Pension Establishment of the United States (1835 Pension Rolls), (Duff Green, Washington, D.C.: 1835).

[66] *Invalid Pensioners*, Letter from the Secretary of the Interior, 31st Congress, 1st Session, House of Representatives, Ex. Doc. No. 74, 22 July 1850.

[67] White, Virgil D., *Index to War of 1812 Pension Files, Volume I (A-I) and Volume II (J-Z)*, (National Historical Publishing Company: Waynesboro, Tennessee, 1992).

[68] White, Virgil D., *Index to U.S. Military Pension Applications of Remarried Widows for Service Between 1812 & 1911*, (National Historical Publishing Company, Waynesboro, Tennessee: 1999).

[69] Scott, Craig R., *The "Lost" Pensions*, Settled Accounts of the Act of 6 April 1838, (Willow Bend Books, Lovettsville, Virginia: 1996).

*The "Lost" Pensions* lists the settled pension accounts prior to 1838. Many survivors simply had not applied for a pension for any of the earlier wars of the United States and Congress wanted to settle these accounts. Under the Congressional Act of 6 April 1838, a deadline was established and those who applied for a pension after this date would be forever rejected. Those who applied for a pension under this act received a one-time cash settlement from the government. Cornelia Mason was not listed in this book's index.

Now that I knew the name of Alexander Mason's wife, it was time to revisit the Internet and to try my Boolean search technique again. This time I typed into my browser "Cornelia Mason" and then Alexander

CHAP. CXXX.—*An Act for the relief of Cornelia Mason.*

STATUTE I.

April 20, 1818.

*Be it enacted, &c.,* That the Secretary of War be, and he is hereby, authorized to place on the half-pay pension list, for five years, at the rate of four dollars per month, Cornelia Mason, the widow of Alexander Mason, who volunteered his services as a militia man, with a detachment of militia commanded by Brigadier-General Perkins, on the northern frontier, in the year one thousand eight hundred and twelve, and who was killed in a battle with a party of Indians, in the month of September in the said year, to be to the use of her and her six children, the legitimate offspring of her, the said Cornelia Mason, and her said deceased husband, Alexander Mason, under the rules and regulations prescribed in, and provided for by, an act entitled "An act making further provision for military services during the late war, and for other purposes," approved April sixteenth, one thousand eight hundred and sixteen.

APPROVED, April 20, 1818.

To be placed on the half-pay pension list.

Act of April 16, 1816, ch. 55.

An act for the relief of Cornelia Mason

(without quotes). What was returned to my screen after I hit "enter" started the brick wall to crumble.

Two web pages containing pictures of the Congressional bills for the relief of Cornelia Mason showed up on my screen. These pages came from the Library of Congress website. Cornelia had to petition Congress to have a pension awarded to her for the death of her husband. The first bill, H.R. 165,[70] was read twice on the floor of the House of Representatives on 27 March 1818 and then submitted to a committee. The second bill, H.R. 172,[71] was read twice and sent to a committee on 1 April 1818. Bills needed to be passed and turned into acts (laws) before anything happens in our government. Now to find this act, which gave Cornelia her pension.

I found a copy of the *Journal of the House of Representatives*[72] on the Internet, which outlined the events, which led to the passage of a bill for the relief of Cornelia:

February 10, 1818 – Congressman Peter Hitchcock of Ohio's 6th Congressional District presented the petition of Cornelia Mason, widow of Alexander Mason, deceased, who was killed in the military service of the United States, respectively praying for a pension.

March 27, 1818 – Congressman John Rhea of Tennessee read the bill for Cornelia Mason (H.R. 165) and it was then sent to the Committee of Pensions and Revolutionary Claims.

April 1, 1818 - Congressman John Rhea, who was on the Committee of Pensions and Revolutionary Claims, read the bill again on the floor on the House.

April 4, 1818 - Congressman John Rhea presented this bill to the Senate for their approval.

---

[70] *The Library of Congress*, American Memory, Government, Law, http://memory.loc.gov/ll/llhb/047/0400/04410000.gif

[71] *Ibid*, http://memory.loc.gov/ll/llhb/047/0400/04650000.gif

[72] *Journal of the House of Representatives*, first session, 15th congress, (E. De Krafft: Washington, DC, 1817).

April 17, 1818 – The bill for Cornelia was approved by both the House and the Senate but since the bill was attached to another bill for Landon Carter's pension request, which still had issues, the bills were sent back to committee.

April 18, 1818 – The issues on Carter's bill were resolved and the bill for Cornelia was read for a third time on the floor on the House and it was passed.

April 20, 1818 – Cornelia's bill was passed on the floor of the Senate and it became law after the President signed it.

A second book that I found on the Internet, *The Public Statues at Large of the United States of America,*[73] lists "an act for the relief of Cornelia Mason." The act states that Alexander Mason volunteered his services as a militiaman and that he was killed by Indians while in the service of the United States. Cornelia received her pension under the Pension Act of 16 April 1816.

Little did she know but Melissa was about to finished tearing down the rest of her brick wall. Melissa supplied me with family information on the Mason's. This included the obituary of Cornelia Mason and a family history on the Mason Family.

Alexander Mason was born on 21 October 1778 in Warren, Rhode Island, to Joseph Mason and Lovina Rounds.[74] He married Cornelia Marvin on 27 May 1801 in Herkimer County, New York. She was born on 11 November 1780 and she died on 19 December 1859. She was the daughter of Matthew Marvin and Gertrude Geiwryck. Alexander and Cornelia had six children.

Cornelia's obituary was printed in *The Firelands Pioneer* in 1898.[75] The obituary had many family discrepancies which made me want to continue my research. Two items, which did not fit with the family information that was given to me from Melissa, was that the family came from Canada and that after the fall of Fort Detroit, Alexander took his family to safety in Chester County, Pennsylvania.

The time between the fall of Fort Detroit and the Battle of Marblehead Peninsula is a little over a month, so I doubt that Alexander had time to take his family to safety in eastern Pennsylvania and then make it back to Ohio in time for this battle. It was now time to see if there was anything else in *The Firelands Pioneer* magazine on this family.

In the March 1859 issue of *The Firelands Pioneer,*[76] Cornelia had written a five-page story on the plight of the family during the War of 1812 and on her life after the war. This first-hand account finally solved all of the family mysteries.

Alexander and Cornelia came to Ohio in May 1812 with five of their children from Hollowell Township, Prince Edward County, Upper Canada (now Ontario). They had been living in Canada since 1807. They arrived in Avery Township, Huron County, Ohio, in June. The area of settlement is now in Erie County.

On 16 August 1812, the American army at Fort Detroit surrendered to the British and most of the frontier areas of Ohio were evacuated due to the threat of Indian and British invasions. Alexander took his family to safety to Chester Township in Geauga County, Ohio, and then he returned to protect his farm. He was probably back in Huron County by the first of September.

---

[73] *The Public Statues at Large of the United States of America*, volume IV, (Charles C. Little and James Brown: Boston, Massachusetts, 1846), 15th congress, 1st session, chapter 130, An act for the relief of Cornelia Mason, 20 April 1818.

[74] Mason, Alverdo Hayward, *Genealogy of the Sampson Mason Family*, pp. 95, 165-166, (Alverdo Hayward Mason: East Braintree, Massachusetts, 1902).

[75] Obituary of Cornelia Mason, *The Firelands Pioneer*, volume XI, October 1898, pp. 343-344, The Firelands Historical Society, Norwalk, Ohio.

[76] Fire Lands Reminiscences by Mrs. Cornelia Mason, *The Fire Lands Pioneer*, volume 1, March 1859, pp. 42-46, The Fire Lands Historical Society, Norwalk, Ohio.

Alexander was 34 years old when he was killed and from all appearances, he was able-bodied and fit to be in the militia. By Ohio law, at this time, the local militia captain had twenty days in which to enroll into the militia all able-bodied males of age who had moved into his area.[77] Alexander was probably a member of Captain David Barrett militia company.[78] This was the first militia company formed in Huron County, Ohio.

We know from local histories that Captain Barrett had called his men to duty after the fall of Fort Detroit. Most of his men had probably left the area with their families so the company did not have its full complement of militiamen. Once General Perkins arrived in Huron County with his brigade, Captain Barrett's company was relieved and the men were told to return to their families. Since Captain Barrett's company was not called to duty by the state or by the federal government, there would be no muster roll or payroll report for his company, thus, no military service records for Alexander Mason at the National Archives.

With the brick wall now in rubble, Melissa asked me for one last favor, that is, to help find the gravesite of Cornelia Mason. Although, Cornelia had died in Tiffin, she was supposed to have been buried in Richland County, Ohio. With a little simple research, I found that Cornelia was buried in the Mansfield Cemetery, which is across the road from the Ohio Genealogical Society's library. This case study had gone full circle.

With the brick wall down, Melissa was able to complete her application and she is now a member of the *National Society United States Daughters of 1812*. As for myself, I have added valuable information to my database on this battle and I can continue researching the men who participated in Ohio's First Battle during the War of 1812.

Some side notes:

Cornelia Mason gave birth to her sixth child, Alexander, Junior, on 8 March 1813. This is five months and seven days after the death of her husband. This is Melissa Danielsson's direct line. Cornelia was pregnant at the time of her husband's death.

Congressman Joshua R. Giddings of Ohio's 20th Congressional District petitioned the House of Representative for a pension for Cornelia on 28 December 1846[79] and another petition for a pension on 31 January 1851.[80] Both petitions failed. It is not know why he did this since she was entitled to only one pension

Giddings happens to be a survivor of the Battle of Marblehead Peninsula and he was probably with Alexander when he was killed. Originally, Congressman Giddings threw me off of my research since I had tried to find Cornelia's act in 1846 and again in 1851. Finding the right House's journal on the Internet finally solved this issue.

---

[77] *Acts Passed at the First Session of the Seventh General Assembly of the State of Ohio, begun and held in the Town of Chillicothe, December 5, 1808, and in the Seventh Year of the Said State*, (J. S. Collins, Chillicothe, Ohio: 1809).

[78] *The Fire Lands Pioneer*, March 1859, (Sandusky, Ohio), Reminiscences of the Hon. F. W. Fowler of Milan, <u>First Organization of the Militia, and Incidents</u>, pp. 2-3, The Fire Lands Historical Society, Norwalk, Ohio.

[79] *Journal of the House of Representatives*, second session, 31st congress, (Washington, DC, 1850-51).

[80] *Ibid*, second session, 29th congress, (Ritchie & Heiss: Washington, DC, 1846-47).

# The Fighting McDonalds of Ross County

During the War of 1812 Ross County, Ohio, produced a family of fighting men who served this nation with distinction in both the regular army and with the Ohio militia. Although not as well know as the Civil War's Fighting McCooks of Carroll County, Ohio, Thomas, James, William and John McDonald made their marks in the history of both Ohio and this nation by becoming prominent military officers.

The four men were the sons of William McDonald and Effie McDonald, originally from the County of Sutherland, Scotland. William was born in 1745 and he died on 6 September 1823 in Ross County. He was the son of Thomas McDonald and Henrietta Gray. William came to America in 1772 following his brother John who arrived in 1770. During the Revolutionary War, William served in the Northumberland County, Pennsylvania, Militia.

William married Effie McDonald in 1774 in Northumberland County. Effie was the daughter of William McDonald and Elizabeth Douglas of the County of Sutherland. She was born in 1730 and she died on 9 September 1823 in Chillicothe, Ross County. Besides John, Thomas, James and William, the family included Nancy, Henrietta and Hiram.

The family moved to the Northwest Territory in 1780 and settled near what is now Steubenville in Jefferson County, Ohio. In 1789 then moved to Mason County, Kentucky, and in 1792 to the Massie's Station on the Ohio side of the Ohio River. In 1796, the family settled in the newly laid out village of Chillicothe.

## Colonel John McDonald

John McDonald was born on 28 January 1775 in Northumberland County and he died on 4 September 1853 at Poplar Ridge in Ross County. He married Catherine Cutright on 5 February 1799 in Ross County. She was the daughter of John Cutright and Elizabeth Subre. Catherine was born in Virginia on 19 August 1750 and she died on 22 March 1850 at Poplar Ridge.

John and Catherine had Effie McDonald (1801-1817), Maria McDonald (1802-1820), Henrietta McDonald (1804-    ), Nancy McDonald (1806-    ), Mary Teter McDonald (1808-    ), John Cutright McDonald (1809-1985), and Margaret McDonald (1811-1814).

John and his brother Thomas served as spies in Major General Anthony Wayne's army during the Indian Wars of 1790's. He had worked as a boatman on the Ohio River, a hunter, and a chain carrier with the early surveyors before moving to Chillicothe.

Before the War of 1812 he worked his way up to lieutenant colonel in the militia. He served as a paymaster and quartermaster on his brother-in-law Colonel Duncan McArthur's regimental staff. McArthur's regiment served with General Hull's army at Detroit and John became a prisoner of war when Hull surrendered his army on 16 August 1812.

John was commissioned a captain in the U.S. Army and served in the 26th Regiment of U.S. Infantry between 20 May 1813 and 7 August 1813. After resigning from the army, John, now a colonel in the militia, commanded a regiment at Fort Detroit and served there until the end of the war. After the war, he would serve two terms in the Ohio Senate.

## Captain Thomas McDonald

Thomas McDonald was born around 1777 in Northumberland County and he died in 1817. He had married a daughter of Samuel Marshall about 1798 in Pennsylvania. Thomas and his brother John served as spies in Major General Anthony Wayne's army during the Indian Wars of 1790's. Later, Thomas would become a captain in the militia, a justice of the peace and a member of the Ohio Assembly.

## Colonel James McDonald

James McDonald was born in 1782 in Northumberland County and he died in 1827. He was commissioned a captain in the Regiment of Rifles, regular army, on 3 May 1808. He was promoted to major on 1 August 1812 and then to lieutenant colonel on 24 January 1814. During the Battle of Fort Erie, Upper Canada, he was brevetted to the rank of colonel on 17 September 1814 for distinguished and meritorious conduct during a sortie from the fort.

On 17 September 1814 he was promoted to colonel and given the command of the newly formed 4th Regiment of U.S. Rifles. At war's end, he was transferred to the 7th Regiment of U.S. Infantry on 17 May 1815 and he resigned from the army on 30 April 1817. James died in 1830 leaving a wife and children.

## Major William McDonald

William McDonald was born about 1784 in Northumberland County and he died on 14 April 1824 at Stoney Creek in Ross County. He married Mary Ann Willis on 26 April 1808 in Northumberland County. They had Louisa Keets McDonald, George McDonald, Effie McDonald and William McDonald.

William was commissioned as a first lieutenant in the regular army on 20 May 1813 and he was assigned to the 26th Regiment of U.S. Infantry which was being raised in western Ohio. He was transferred to the 19th Regiment of U.S. Infantry on 12 May 1814 and he was promoted to captain on 11 November 1814.

William was brevetted major on 25 July 1814 for distinguished service at the Battle of Lundy's Lane, Upper Canada. At war's end, he was transferred to the 3rd Regiment of U.S. Infantry on 17 May 1815. He served as an assistant inspector general holding rank of major from 29 April 1816 to 17 October 1820. His rank was reverted to captain on 20 May 1820 and he was dismissed from the army on 17 October 1820. After the war he would become a sheriff and later a deputy marshal for the state.

## Nancy McDonald

Nancy McDonald was born in 1779 in Northumberland County and she died on 23 October 1836 in Chillicothe. She married Duncan McArthur in February 1796 in Ross County. Duncan was a major general in the Ohio militia when war broke out in 1812 and he became a colonel in Brigadier General William Hull's Army of the Northwest which surrendered to the British in August 1812.

Duncan was commissioned as a colonel in the regular army and was promoted to brigadier general. He would command the Army of the Northwest after Major General William Henry Harrison resigned from the army in 1814. After the war, he became a politician and served as the governor of Ohio.

## Henrietta McDonald

Henrietta McDonald was born in 1790 in Ross County and she died between 1830 and 1837 in Ross County. She married Pressley Morris in 1811 in Ross County.

## Hiram McDonald

Hiram McDonald was born about 1792 in Ross County and died young in 1796 in Ross County.

## Bibliography

Evans, Lyle S., *A Standard History of Ross County, Ohio*, volume II, (Chicago and New York: The Lewis Publishing Company, 1918), pp. 575-582, Colonel John McDonald.

Heitman, Francis B., *Historical Register and Dictionary of the United States Army From Its Organization, September 29, 1789, to March 2, 1903*, Volume I, (Genealogical Publishing Company, Baltimore, Maryland: 1994), part II, complete alphabetical list of all commissioned officers, pp. 662-663, McDonalds.

Taylor, William A., *Ohio Statesmen and Annals of Progress from the Year 1788 to the Year 1900*, volume 1, (Columbus, OH: The Westbote Co., State Printers, 1899), pp. 85

*Roster of Ohio Soldiers in the War of 1812*, The Adjutant General of Ohio 1816, (Heritage Books, Inc., Bowie, Maryland: 1995), Roll of Field and Staff, Colonel Duncan McArthur's First Regiment, Ohio Militia, page 5.

# Ohio's first graduates of West Point

Ohio's first graduates of the U.S. Military Academy at West Point, New York, were in the Class of 1815 barely missing the War of 1812.

John R. Sloo, cadet number 129, started classes at West Point on 25 June 1813 and he was commissioned a third lieutenant of artillery on 2 March 1815 less than a month after the Treaty of Ghent was ratified.

Sloo was born in Kentucky and received his appointed from West Point while living in Ohio. He would serve at Fort Mifflin, outside of Philadelphia, Pennsylvania, from 1815 to 1818. He resigned from the army on 30 April 1818 after being promoted to second lieutenant. After leaving the military, Sloo moved to Shawneetown, Illinois, where he was the register of the U.S. Land Office. He died in 1837.

Henry W. Griswold, cadet number 130, started his military career at West Point on 28 July 1813 and he also was commissioned as a third lieutenant of artillery on 2 March 1815.

Griswold was first Ohio born graduate of West Point and he made the military his career. He served at Fort Niagara and Sackets Harbor, New York; Fort Washington, Maryland; in New York harbor; on commissary duty; as an assistant instructor at West Point; Fort Independent, Massachusetts; Fort Monroe, Virginia; and Fort Mitchell, Alabama. He died on 23 October 1834 at Castle Pinckney, South Carolina. When he died he was a captain in the 3rd Artillery Regiment.

Aaron G. Gano, cadet number 143, started classes on 8 January 1814 later than the first two cadets but he was commissioned as a third lieutenant of artillery on 2 March 1815 along with Sloo and Griswold.

Ohio born, Gano was the son of Major General John S. Gano of the Ohio Militia. He served in garrison in New York harbor from 1815 to 1817. He resigned from the army on 1 October 1817. Gano became a merchant in Cincinnati, Ohio, between 1817 and 1840 and then at Hannibal, Missouri, between 1840 and 1854. He died near Cincinnati on 2 December 1854 at the age of 58.

James R. Stubbs, cadet number 158, started classes with Gano on 8 January 1814 and he was commissioned on 11 December 1815 as a brevet second lieutenant in the light artillery.

Also Ohio born, Stubbs would serve in various New England posts between 1815 and 1817 before he resigned from the army on 15 November 1817. He was reappointed as a captain on 30 November 1819 in the Quartermaster Corps and worked there until the corps was disbanded on 1 June 1821. Stubbs then worked as a clerk in the Post Office Department in Washington, D.C., between 1823 and 1829. He died in 1832 while in Cincinnati.

The next Ohioan to graduate from West Point was Thomas McArthur in the Class of 1820. He was the son of Brigadier General Duncan McArthur, U.S. Army, and later a governor of Ohio. Thomas began classes at West Point on 5 April 1816 and he was commissioned on 1 July 1820. He was authorized a leave of absence upon graduation and he resigned from the army on 24 October 1820 without having served. He would become a merchant in Springfield, Ohio, and he died on 21 February 1833 at the age of 31 while in Chillicothe.

# Forgotten Companies

Forgotten Companies

## Discovering the truth about the fall of Fort Sullivan

Fort Sullivan in Eastport, Maine, fell to the British on 11 July 1814 without a shot being fired. The commanding officer, Major Perley Putnam of the 40th Regiment of U.S. Infantry, surrendered the fort when faced with overwhelming odds and no chance of a victory.

Benson Lossing in this book, *The Pictorial Field-Book of the War of 1812*, states that Putnam had 50 soldiers, six artillery pieces and 250 local militiamen under his command. The British came with the ship-of-the-line *Ramillies*, the sloop *Martin*, the brig *Bream*, the bomb-ship *Terror* and two troopships filled with British soldiers. Every book on the War of 1812 which mentions this minor incident has simply repeated Lossing's account without doing any further investigations.

Major Putnam wrote an after-action report on 8 August 1814 listing the 83 men of the 40th Infantry, including himself, who surrendered to the British the previous month. In this list are the names of two captains, a second lieutenant and two ensigns. This seems very odd that there were two captains and no first lieutenants and no third lieutenants assigned to his command. There should have been slightly over 200 enlisted men with ten officers serving under Putman.

Of the 83 men, six officers and 37 men were released as prisoners of war on parole status and sent home. These men could not take up arms against Great Britain until they were listed on a prisoner of war exchange list and then exchanged for a British soldier. The rest of the men were listed as prisoners of war at Eastport.

While researching the American prisoner of war records for the British prison camp at Halifax, Nova Scotia, I came across 86 men who were listed as members of the U.S. Sea Fencibles. The U.S. Army raised ten companies of sea fencibles of which the nearest company to Eastport was raised in Portsmouth, New Hampshire. This company served at Fort Independence in Portsmouth and there were no known U.S. Sea Fencibles companies captured by the British during this war.

Some of the men with the sea fencibles at Halifax were listed as privates while other men were listed as seamen. When I started to research each of the privates, using the *U.S. Army Register of Enlistments*, I found that 55 of these men were members of the 40th Infantry who were captured by the HMS *Ramillies* on 11 July 1814. Furthermore, of the 31 remaining men held at Halifax, two were masters, two were mates and the rest were seamen. There were also two civilians listed with the sea fencibles. A master is a captain of a civilian ship and there were two of them on this list. Another mystery to investigate!

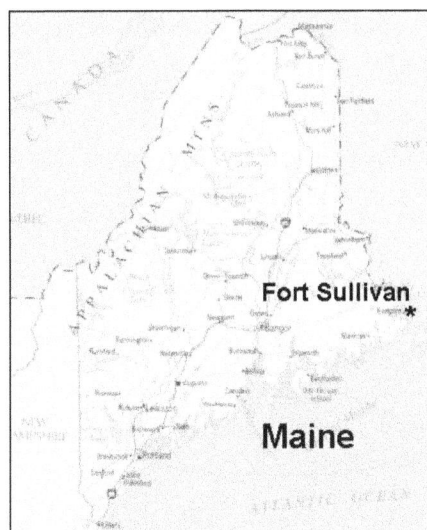

Fort Sullivan
Eastport, District of Maine

It was time to create a database using Putnam's list and the sea fencibles list from Halifax to see if the additional 55 soldiers captured by the *Ramillies* were listed on Putnam's report. Most of the names matched between the two reports revealing only seven men that Putman had forgotten to add to his list. A total of ninety men had surrendered to the British and I needed to find another 100 or so men to solve this mystery.

It was now time to research each of the 90 men using the *Register of Enlistments* plus the land bounty records and the pension records for this war. Ensign Isaac Carpenter's affidavit in his pension records solved most of the mysteries surrounding the surrendering of Fort Sullivan. He was a part of a detachment of the 40th Infantry which had left Boston and had marched across Maine for thirty days in very bad weather. Many of the men were in bad physical condition and along with Ensign Carpenter, who was sick, they were put on board a schooner bound for Eastport. The ensign gives no number of how many men were in this detachment or how many men were transported on the schooner. Regardless, these men were captured by the British before the fort surrendered.

It appears now that Major Putman had arrived early to Eastport with his staff and some guards to take command of Fort Sullivan. The first detachment of soldiers was on the march from Boston and a second detachment was probably being organized when the fort fell. The 100 or so missing men were most likely still in Boston, the headquarters for the 40th Infantry.

It appears that the British actually captured two schooners with the sick men from the 40th Infantry. Each schooner would have had a master in command. The rest of the detachment was still enroute to the fort and were in no condition to defend this facility. Between the fort and the British invasion force was the village of Eastport. The militiamen probably forced Putnam to surrender, for good reasons, their homes and families were in the way of a potential battle.

Eastport sits on Moose Island in Passamaquoddy Bay between Maine and New Brunswick, Canada. It had been claimed by Great Britain since the end of the Revolutionary War. After the fall of Eastport, the British went on to capture the villages of Hampton, Bangor and Machias. With approximately two-thirds of Maine under their control, the British organized the colony of New Ireland. At the end of the war the British returned all of captured Maine, except Eastport, to the United States. They would return this village and the island to the United States in 1818.

Fort Sullivan was the eastern most land 'battle' fought during the War of 1812 in the United States. The return of Eastport and Moose Island marks the true end of the War of 1812.

## Soldiers of the 40th U.S. Infantry captured at Fort Sullivan

Alston, Daniel A. - Private - Company: John Fillebrown – Prisoner of War who was sent to Fort Wolcott.

Barfield, Richard - Fifer - Company: John Fillebrown - Age: 40 - Height: 5' 6" - Born: Manchester, Essex County, MA - Trade: Scrivener (Scribe) - Enlistment date: 27 Dec 1813 - Enlistment place: Watertown - Enlistment period: War - By whom: Lieutenant Blanchford - Captured and escaped, at Machias, ME.

Bird Jr., William - Sergeant - Company: Jacob Varnum - Age: 28 - Height: 5' 10 1/2" - Born: Boston, MA - Trade: Hatter - Enlistment date: 18 Feb 1814 - Enlistment place: Boston, MA - Enlistment period: War - Bounty: BLW 12770-160-12 - Captured and escaped, at principle rendezvous at Boston; discharged at Boston on 5 Apr 1815.

Blanchard, Reuben K. - Second Lieutenant - Company: John Fillebrown - Bounty: BLW 10000-170-50 - Commissioned as a 2nd lieutenant, 40th Infantry, on 1 Sep 1813; on parole of honor at Charlestown, MA; promoted to 1st lieutenant on 12 Oct 1814; discharged on 15 Jun 1815.

Brown, John - Private - Company: Jacob Varnum - Age: 27 - Height: 5' 11 1/2" - Born: Kingston, NH - Trade: Shoemaker - Enlistment date: 6 Jan 1814 - Enlistment place: Salem, MA - Enlistment period: War - By whom: Lieutenant Manning - Bounty: BLW 16417-160-12 - Prisoner of War at Eastport, ME; discharged on 5 Apr 1815.

Caldwell, John - Sergeant - Company: Jacob Varnum - Age: 30 - Height: 5' 9" - Born: Ware, Hampshire County, MA - Trade: Farmer - Enlistment date: 30 Oct 1813 - Enlistment place: Concord, NH - Enlistment period: War - By whom: Capt Leonard - Bounty: BLW 5532-160-12 - Captured and escaped, at Machias, ME; discharged at Castle Island on 18 Oct 1815.

Carpenter, Isaac - Third Lieutenant - Company: Jacob Varnum - Pension: Old War Invalid 27592 - Bounty: BLW 6259-160-50 - Commissioned as an ensign, 40th Infantry, on 1 Sep 1813; promoted to 3rd lieutenant on 2 May 1814; on parole of honor at Western; discharged on 15 Jun 1815.

Carr, John - Private - Company: Jacob Varnum - Age: 21 - Height: 5' 5" - Born: New Durham, NH - Trade: Blacksmith - Enlistment date: 21 Feb 1814 - Enlistment place: Concord, NH - Enlistment period: War - By whom: Capt Leonard - Bounty: BLW 14485-160-12 - Prisoner of War at Eastport, ME;

discharged at Boston on 5 Apr 1815.

Chapman, James - Private - Age: 32 - Height: 5' 9 1/2" - Born: Nobleboro, Lincoln County, MA - Trade: Farmer - Enlistment date: 16 Feb 1814 - Enlistment place: Hallowell - Enlistment period: War - By whom: Capt Fillebrown - Bounty: BLW 10395-160-12 - On parole at Machias, ME; discharged at Boston on 5 Apr 1815.

Copps, Darius - Ensign - Commissioned as an ensign, 40th Infantry, on 2 Mar 1814; on parole, since discharged from the service; struck off on 4 Jul 1814.

Crawford, Thomas - Private - Company: Jacob Varnum - Age: 26 - Height: 5' 9" - Born: Bath, MA - Trade: Farmer - Enlistment date: 9 Apr 1814 - Enlistment place: Hallowell - Enlistment period: War - By whom: Fillebrown - Bounty: BLW 12632-160-12 - Prisoner of War at Eastport; Discharged at Boston on 5 Apr 1815.

Creach, Thomas B. - Sergeant - Company: Jacob Varnum - Age: 28 - Height: 5' 9" - Born: Plymouth, MA - Trade: Carpenter - Enlistment date: 14 Feb 1814 - Enlistment place: Dedham, MA - Enlistment period: War - By whom: 1st Lieutenant Durant - Bounty: BLW 8463-160-12 - Captured and escaped, at Hallowell; promoted to sergeant on 1 Mar 1815; discharged at Boston on 5 Apr 1815.

Dailey, John - Private - Company: Jacob Varnum - Age: 23 - Height: 5' 9 1/2" - Born: Woodstock, Windham County, CT - Trade: Laborer - Enlistment date: 17 Mar 1814 - Enlistment place: Boston, MA - Enlistment period: War - By whom: Capt Nye - Prisoner of War at Eastport; Rose Island, Fort Walcott, Halifax.

Dell, John - 40th Infantry.

Donaldson, Frederick - Private - Company: Jacob Varnum - Bounty: BLW 11732-160-12 - On parole at Machias.

Dow, Frederick - Private - Company: Jacob Varnum - Age: 23 - Height: 5' 11" - Born: Hopkinton, NH - Trade: Farmer - Enlistment date: 28 Feb 1814 - Enlistment place: Concord, NH - Enlistment period: War - By whom: Capt. Leonard - Pension: Wife Sabra A. Pratt, WO-12926,WC-14366; married on 10 Jun 1828 in Pernette Square, Jefferson County, NY; soldier died on 5 Dec 1865 in Brighton, Lorain County, Ohio; wife died on 25 Jun 1883 - Bounty: BLW 6203-160-12 - Prisoner of War at Eastport; discharged at Boston on 5 Apr 1815.

Duntin, Moody - Private - Company: Jacob Varnum - Age: 27 - Height: 5' 11" - Born: New Sales, Rockingham, NH - Trade: Carpenter - Enlistment date: 27 Dec 1813 - Enlistment place: Concord, NH - Enlistment period: War - By whom: Capt Leonard -Prisoner of War at Eastport; discharged on 5 Apr 1815.

Dunton, Peleg - Private - Company: Jacob Varnum - Age: 18 - Height: 5' 7 1/2" - Born: Marlborough, Middlesex, MA - Trade: Laborer - Enlistment date: 31 Dec 1813 - Enlistment period: War - By whom: Varnum - Bounty: BLW 21085-160-12 - Prisoner of War at Eastport; sent to Halifax; sent to Fort Wolcott; Ross Island; discharged at New London on 20 Aug 1815.

Eaton, Hubbard - Private - Company: John Fillebrown - Age: 20 - Height: 5' 8 1/2" - Born: Monmouth, ME - Trade: Farmer - Enlistment date: 21 Feb 1814 - Enlistment place: Hallowell - Enlistment period: War - By whom: Capt Fillebrown - Bounty: BLW 6304-160-12 - Prisoner of War at Eastport; sent to Halifax; sent to Fort Wolcott; discharged at New London on 20 Aug 1815.

Egans, James - Private - Company: John Fillebrown - Age: 19 - Height: 5' 5 1/4" - Born: Ipswich, MA - Trade: Blacksmith - Enlistment date: 21 Feb 1814 - Enlistment place: Boston, MA - Enlistment period: War - By whom: Lieutenant Blanchard - Bounty: BLW 1355-160-12 - Prisoner of War at Eastport; discharged at New London on 20 Aug 1815.

Erskine, Robert - Corporal - Company: John Fillebrown - Age: 22 - Height: 5 11 1/2" - Born: Bristol, MA - Trade: Farmer - Enlistment date: 1 Mar 1814 - Enlistment place: Hallowell - Enlistment period: War - Bounty: BLW 16271-160-12 - Captured and escaped, at principle rendezvous at Boston; discharged at Boston on 21 Mar 1815.

Eveleth, Benjamin W. - Private - Company: John Fillebrown - Age: 21 - Height: 5' 9 1/2" - Born: Gardner, ME - Trade: Farmer - Enlistment date: 17 Mar 1814 - Enlistment place: Hallowell - Enlistment period: War - By whom: Capt Fillebrown - Bounty: BLW 19234-160-12 - Land bounty to James Eveleth, father and heir at law of Benjamin W. Eveleth - Prisoner of War at Eastport; set to Halifax; died at Halifax on 1 Oct 1814 from fever.

Favour, John - Private - Company: Jacob Varnum - Age: 32 - Height: 5' 10" - Born: Amesbury, MA - Trade: Blacksmith - Enlistment date: 9 Feb 1814 - Enlistment place: Concord, NH - Enlistment period: War - Bounty: BLW 6718-160-12 - On parole at Machias; discharged at Boston on 5 Apr 1815.

Fillebrown Jr., John - Captain - Company: John Fillebrown - Commissioned as a captain, 40th Infantry, on 1 Sep 1813; on parole of honor at Boston; discharged on 15 Jun 1815.

Fillebrown, Richard - Sergeant - Company: Jacob Varnum - Other regiment: ex 4th Infantry - Age: 30 - Height: 5' 6 1/2" - Born: Cambridge, MA - Trade: Blacksmith - Enlistment date: 15 Oct 1808 - Enlistment place: Fuller - Enlistment period: 5 Yrs - Battle of Tippecanoe, Captain Josiah Snelling's Company, sergeant - Bounty: BLW 8502-160-12 - Prisoner of War (probably at Detroit. MI), furloughed under General Orders on 11 Dec 1812; return of United States troops who being prisoners of war to the British government and sent on parole from Quebec, have been landed on Castle Island in the harbor of Boston between the dates of November 24th and November 30th 1812; discharged on 15 Oct 1813; re-enlisted in the 40th Infantry at Taunton or Dedham on 19 Mar 1814 for the war; captured and escaped, at principle rendezvous at Boston; discharged on 5 Apr 1815.

Fillebrown, Samuel S. - Sergeant - Company: Jacob Varnum - Age: 33 - Height: 5' 6" - Born: Cambridge, MA - Trade: Carpenter - Enlistment date: 14 Feb 1814 - Enlistment place: Dedham, MA - Enlistment period: War - By whom: Durant - Pension: Wife Dorcas Hayes, Old War Widows File 13369 - Bounty: BLW 8475-160-12 - Prisoner of War at Eastport and Halifax; discharged at Boston on 5 Apr 1815.

Foss, Richard - Private - Company: John Fillebrown - Age: 23 - Height: 5' 8" - Born: Saco, ME - Trade: Farmer - Enlistment date: 22 Feb 1814 - Enlistment place: Hallowell - Enlistment period: War - By whom: Fillebrown - Bounty: BLW 5824-160-12 - Prisoner of War at Eastport; sent to Halifax; sent to Fort Wolcott; discharged at New London, CT, on 20 Aug 1815.

Foster, Edla - Corporal - Company: Jacob Varnum - Age: 24 - Height: 5' 10 1/2" - Born: Dunbarton, NH - Trade: Carpenter - Enlistment date: 5 Jan 1814 - Enlistment place: Concord, NH - Enlistment period: War - Bounty: BLW 8490-160-12 - Captured and escaped, at principle rendezvous at Boston.

Foster, William - Private - Company: Jacob Varnum - Age: 21 - Height: 5' 11" - Born: Alexandria, DC - Trade: Farmer - Enlistment date: 28 Feb 1814 - Enlistment place: Concord, NH - Enlistment period: War - By whom: Lieutenant Hodges - Bounty: BLW 13491-160-12 - Prisoner of War at Eastport; discharged at Boston on 5 Apr 1815.

Gahan, Jeremiah - Private - Company: John Fillebrown - Age: 24 - Height: 5' 7" - Born: Bethel, ME - Trade: Farmer - Enlistment date: 16 Feb 1814 - Enlistment place: Hallowell - Enlistment period: War - By whom: Fillebrown - Bounty: BLW 7330-160-12 - Prisoner of War at Easton; Discharged at Boston on 5 Apr 1815.

Goodwin, Robert - Private - Company: John Fillebrown - Age: 24 - Height: 5' 8" - Born: Dresden, MA - Trade: Mariner - Enlistment date: 26 Feb 1814 - Enlistment place: Hallowell - Enlistment period: War - Pension: Wife Elvira D., SO-7891, SC-18350, WO-35074, WC-21900; married on 2 Mar 1861 in New London, CT (1st wife Eunice Goffery); soldier died on 18 Feb 1879 in New London, CT; widow died on 14 May 1886 in Waterford, CT - Bounty: BLW 3548-160-12 - Prisoner of War at Halifax; sent to Fort Wolcott; discharged at New London on 20 Aug 1815.

Gott, Andrew - Private - Company: Jacob Varnum - Age: 17 - Height: 5' 6 1/2" - Born: New Boston, NH - Trade: Farmer - Enlistment date: 22 Nov 1813 - Enlistment place: Concord, NH - Enlistment period: War - Bounty: BLW 11031-160-12 - Prisoner of War at Eastport; discharged at Boston on 5 Apr 1815.

Graham, Jesse - 40th Infantry.

Hall, Silas - Private - Age: 21 - Height: 5' 7" - Born: Cumberland, MA - Trade: Farmer - Enlistment date: 21 Mar 1814 - Enlistment place: Fairhaven, MA - Enlistment period: War - By whom: Lieutenant McComb - Pension: Wife Hannah, SO-9936, SC-5453, WO-23475, WC-23247 - Bounty: BLW 7332-160-12 - On parole at Machias; discharged on 10 Apr 1815.

Hamilton, John - Private - Company: John Fillebrown - Age: 21 - Height: 5' 10 1/2" - Born: Augusta, ME - Trade: Farmer - Enlistment date: 18 Mar 1814 - Enlistment place: Hallowell - Enlistment period: War - By whom: Fillebrown - Bounty: BLW 16273-160-12 - Captured and escaped, at Hallowell; discharged at Boston on 31 Mar 1815.

Hanscom, Eleazer - Private - Age: 30 - Height: 5' 9 1/2" - Born: Cape Elizabeth, MA - Trade: Shoemaker - Enlistment date: 19 Feb 1814 - Enlistment place: Hallowell - Enlistment period: War - By whom: Fillebrown - Prisoner of War at Eastport; discharged at Boston on 5 Apr 1815.

Hanson, Samuel - Private - Company: Jacob Varnum - Age: 30 - Height: 5' 7" - Born: Berwick, ME - Trade: Mariner - Enlistment date: 22 Oct 1813 - Enlistment place: Boston, MA - Enlistment period: War - By whom: Nye - Bounty: BLW 6194-160-12 - Prisoner of War at Eastport; discharged from Castle Island on 13 Sep 1815.

Hardy, Reuben - Private - Company: Jacob Varnum - Age: 33 - Height: 5' 7" - Born: Nottingham, NH - Trade: Farmer - Enlistment date: 28 Feb 1814 - Enlistment place: Concord, NH - Enlistment period: War - Bounty: BLW 8726-160-12 - On parole at Machias; discharged at Portland, ME, on 10 Apr 1815.

Harrington, Jason - Private - Company: John Fillebrown - Age: 30 - Height: 5' 7 1/2' - Born: Orange, MA - Trade: Laborer - Enlistment date: 26 Feb 1814 - Enlistment place: Boston, MA - Enlistment period: War - By whom: Lieutenant Blanchard - Bounty: BLW 11744-160-12 - Prisoner of War at Easton; sent to Halifax; sent to Fort Wolcott; discharged at New London, CT, on 20 Aug 1815.

Harvey, Samuel - 40th Infantry.

Haskell, William - Private - Company: John Fillebrown - Age: 21 - Height: 5' 4 1/2" - Born: Falmouth,

MA - Trade: Farmer - Enlistment date: 4 Mar 1814 - Enlistment place: Hallowell - Enlistment period: War - By whom: Fillebrown - Pension: Wife Betsey McMaster, Old War Widows File 12447 - Bounty: Heirs received half pay for five years in lieu of land bounty (William and Greenleaf Haskell, Kennebec County, ME) - On parole at Salem; died at the regimental rendezvous on 31 Mar 1815, buried at the arsenal.

Henderson, John - Private - Company: Jacob Varnum - Age: 42 - Height: 5' 5" - Born: Hopkinton, NH - Trade: Farmer - Enlistment date: 7 Sep 1813 - Enlistment place: Concord, NH - Enlistment period: War - By whom: John Leonard - Bounty: BLW 5344-160-12 - Prisoner of War at Eastport; sent to Halifax; sent to Fort Wolcott; discharged at New London, CT, on 20 Aug 1815.

Higgins, James - Private - Company: John Fillebrown - Prisoner of War on parole, sent to Fort Walcott.

Hill, Jeremiah - Private - Company: John Fillebrown - Age: 21 - Height: 5' 8" - Born: Buxton, MA - Trade: Farmer - Enlistment date: 16 Feb 1814 - Enlistment place: Hallowell - Enlistment period: War - By whom: Fillebrown - Pension: SO-29390, SC-21693 - Bounty: BLW 15292-160-12 - On parole at Salem; discharged at New London, CT, on 20 Aug 1815.

Hogg, Elijah - Private - Company: Jacob Varnum - Age: 23 - Height: 5' 7" - Born: Londonderry, NH - Trade: Farmer - Enlistment date: 13 Oct 1813 - Enlistment place: Concord, NH - Enlistment period: War - By whom: Leonard - Bounty: No bounty found - Prisoner of War at Eastport; discharged at Boston on 5 Apr 1815.

Hutchinson, Joseph - Private - Company: Jacob Varnum - Age: 18 - Height: 5' 7" - Born: Allentown, NH - Trade: Laborer - Enlistment date: 24 Jan 1814 - Enlistment place: Concord, NH - Enlistment period: War - By whom: Leonard - Bounty: BLW 10791-160-12 - Prisoner of War at Eastport; discharged at New London, CT, on 20 Aug 1814.

Hutchinson, Seth - Private - Company: Jacob Varnum - Age: 21 - Height: 6' - Born: Litchfield, ME - Trade: Farmer - Enlistment date: 7 Mar 1814 - Enlistment place: Hallowell - Enlistment period: War - By whom: Fillebrown - On parole at Salem; discharged at Boston on 5 Apr 1815.

Hutchinson, Sewell - Private - Age: 21 - Height: 5' 9" - Born: Litchfield, ME - Trade: Farmer - Enlistment date: 7 Mar 1814 - Enlistment place: Hallowell - Enlistment period: War - By whom: Fillebrown - Bounty: No bounty found - On parole at Salem; set to Halifax; died at Halifax on 17 May 1815.

Hutchinson, William - Private - Company: Jacob Varnum - Age: 28 - Height: 5' 5" - Born: Gilmantown, NH - Trade: Carpenter - Enlistment date: 7 Feb 1814 - Enlistment place: Concord, NH - Enlistment period: War - By whom: Leonard - Bounty: BLW 8489-160-12 - Prisoner of War at Eastport; sent to Halifax; sent to Fort Wolcott; discharged from Fort Independence on 16 Aug 1815.

Kimball, Bradbury - Private - Company: Jacob Varnum - Age: 22 - Height: 6' - Born: Hopkinton, Hillsboro, NH - Trade: Cordwainer - Enlistment date: 27 Dec 1813 - Enlistment place: Concord, NH - Enlistment period: War - Bounty: BLW 5914-160-12 - Prisoner of War at Eastport; discharged on 5 Apr 1815.

Lamb, Oliver - Private - Company: Jacob Varnum - Age: 24 - Height: 5' 10 1/2" - Born: Carlton, MA - Trade: Farmer - Enlistment date: 5 Mar 1814 - Enlistment place: Boston, MA - Enlistment period: War - By whom: Varnum - Bounty: BLW 24734-160-12 - Prisoner of War at Eastport; discharged on 5 Apr 1815.

Lee, George - Private - Company: John Fillebrown - Age: 25 - Height: 5' 11" - Born: Taunton, MA - Trade: Farmer - Enlistment date: 14 Feb 1814 - Enlistment period: War - By whom: Lieutenant Hodges - On parole at Salem; died at Salem, MA, on 3 Apr 1815 from small pox.

Leonard Jr., Samuel - Private - Company: John Fillebrown - Age: 25 - Height: 5' 8" - Born: Taunton, MA - Trade: Laborer - Enlistment date: 19 Nov 1813 - Enlistment place: Taunton, MA - Enlistment period: War - By whom: Lieutenant Hodges - Bounty: BLW 13781-160-12 - On parole at Salem; discharged on 5 Apr 1815.

Leonard, Jacob - Private - Company: John Fillebrown - Age: 18 - Height: 5' 7 1/2" - Born: Taunton, MA - Trade: Saddler - Enlistment date: 13 Jan 1814 - Enlistment place: Taunton, MA - Enlistment period: War - On parole at Salem; discharged on 5 Apr 1815.

Leonard, Samuel - Private - Company: John Fillebrown - Age: 42 - Height: 5' 9 1/2" - Born: Taunton, MA - Trade: Laborer - Enlistment date: 21 Dec 1813 - Enlistment place: Taunton, MA - Enlistment period: War - By whom: Lieutenant Hodges - Bounty: BLW 13781-160-12 - On parole at Salem; sent to Halifax; discharged on 5 Apr 1815.

Lindsay, David - Sergeant - Company: Jacob Varnum - Age: 22 - Height: 5' 9 1/2" - Born: Thornton, NH - Trade: Rope maker - Enlistment date: 20 Feb 1814 - Enlistment place: Watertown - Enlistment period: War - By whom: Varnum - Bounty: BLW 7618-160-12 - Prisoner of War at Eastport; discharged at Boston on 5 Apr 1815.

Mellus, Daniel C. - Private - Company: John Fillebrown - Age: 21 - Height: 5' 7" - Born: Georgetown, Lincoln, ME - Trade: Block maker - Enlistment date: 16 Feb 1814 - Enlistment place: Hallowell - Enlistment period: War - Bounty: BLW 20650-80-50 and 20650-80-50 - On parole at Salem.

Merian, James F. - Private - Company: Jacob Varnum - Age: 21 - Height: 5' 9 1/2" - Born: Charlestown, Middlesex County, MA - Trade: Blacksmith - Enlistment date: 22 Feb 1814 - Enlistment place: Dedham, MA - Enlistment period: War - By whom: Lieutenant Durant - Bounty: BLW 8476-160-12 - Prisoner of War at Eastport; discharged at Boston on 5 Apr 1815.

Merian, Samuel - Private - Company: Jacob Varnum - Age: 18 - Height: 5' 7" - Born: Dorchester, Norfolk, MA - Trade: Brush maker - Enlistment date: 21 Feb 1814 - Enlistment place: Dedham, MA - Enlistment period: War - By whom: Lieutenant Durant - Prisoner of War at Eastport; sent to Fort Wolcott, Rose Island, Halifax; discharged on 17 Jul 1815.

Miller, Philip - Private - Company: John Fillebrown - Age: 21 - Height: 5' 11" - Born: Waldoborough, Lincoln County, ME - Trade: Farmer - Enlistment date: 21 Mar 1814 - Enlistment place: Hallowell - Enlistment period: War - Bounty: BLW 8473-160-12 - On parole at Machias; discharged on 10 Apr 1815.

Mincher, William - Private - Company: Jacob Varnum - Age: 20 - Height: 5' 8" - Born: Lunenbery, Worcester County, MA - Trade: Farmer - Enlistment date: 24 Nov 1813 - Enlistment place: Concord, MA - Enlistment period: War - Prisoner of War at Eastport; discharged on 5 Apr 1815.

Moore, Obadiah - Private - Company: Jacob Varnum - Age: 33 - Height: 5' 7" - Born: Pembroke, NH - Trade: Farmer - Enlistment date: 23 Feb 1814 - Enlistment place: Concord, NH - Enlistment period: War - By whom: Leonard - Bounty: BLW 8497-160-12 - Prisoner of War at Eastport; discharged at Boston on 5 Apr 1815.

Morrison, Moses - Private - Company: Jacob Varnum - Age: 19 - Height: 5' 9" - Born: Candia, NH -

Trade: Farmer - Enlistment date: 17 Feb 1814 - Enlistment place: Charleston - Enlistment period: War - By whom: Third Lieutenant Manning - Bounty: BLW 20273-160-12 - Land bounty to David Morrison, brother and other heirs at law of Moses Morrison - Prisoner of War at Halifax, died at Halifax on 13 Feb 1815 from dysentery.

Mullett, Isaac - Private - Company: Jacob Varnum - Age: 22 - Height: 5' 5 1/2" - Born: Charlestown, Middlesex County, MA - Trade: Painter - Enlistment date: 22 Jan 1814 - Enlistment place: Watertown - Enlistment period: War - By whom: Lieutenant Blanchand - Bounty: BLW 11121-160-12 – On parole at Salem; discharged at Boston on 5 Apr 1815.

Nicholas, Jacob - Private - Company: Jacob Varnum - Age: 40 - Height: 5' 7" - Born: Amsterdam, Holland - Trade: Laborer - Enlistment date: 14 Feb 1814 - Enlistment place: Watertown - Enlistment period: War - Bounty: BLW 8478-160-12 - Captured and escaped, at principle rendezvous at Boston; discharged on 5 Apr 1815.

Nutting, Cyrus - Private - Company: Jacob Varnum - Age: 21 - Height: 5' 9" - Born: Carlisle, MA - Trade: Cordwainer - Enlistment date: 16 Feb 1814 - Enlistment place: Watertown - Enlistment period: War - Bounty: BLW 16048-160-12 - Prisoner of War at Eastport; sent to Fort Wolcott; discharged at New London on 20 Aug 1815.

Nutting, Simri - Private - Company: Jacob Varnum - Age: 22 - Height: 5' 8" - Born: Carlisle, MA - Enlistment date: 14 Mar 1814 - Enlistment period: War - Bounty: BLW 16049-160-12- Prisoner of War at Eastport; sent to Halifax; discharged at New London on 20 Aug 1815.

Orkins, James - Private - Company: Jacob Varnum - Age: 36 - Height: 5' 10" - Born: Kingston, NH - Trade: Farmer - Enlistment date: 14 Feb 1814 - Enlistment place: Concord, NH - Enlistment period: War - By whom: Leonard - Bounty: BLW 5538-160-12 - Prisoner of War at Eastport; discharged at Boston on 5 Apr 1815.

Page, David - Private - Company: Jacob Varnum - Age: 38 - Height: 5' 10 1/4' - Born: Epping, Rockingham, NH - Trade: Joiner - Enlistment date: 18 Feb 1814 - Enlistment place: Hallowell - Enlistment period: War - By whom: Fillebrown - Bounty: BLW 7761-160-12 - On parole at Machias; discharged at Boston on 5 Apr 1815.

Paleifer, Joseph (Pulcifer) - Private - Company: John Fillebrown - Age: 43 - Height: 5' 7" - Born: Cape Ann, Essex County, MA - Trade: Farmer - Enlistment date: 19 Feb 1814 - Enlistment place: Hallowell - Enlistment period: War - By whom: Fillebrown - Bounty: BLW 5481-160-12 - On parole at Salem; at Halifax; at Fort Wolcott; Rose Island; discharged at New London on 20 Apr 1815.

Parker, John - Private - Company: Jacob Varnum - Age: 18 - Height: 5' 4" - Born: Groton, Middlesex, MA - Trade: Laborer - Enlistment date: 26 Nov 1813 - Enlistment place: Concord, MA - Enlistment period: War - Bounty: BLW 13425-160-12 - Captured and escaped, at principle rendezvous at Boston; discharged on 5 Apr 1815.

Paskiel, Ezekiel D. - Private - Age: 21 - Height: 6' 3" - Born: Warren, ME - Trade: Farmer - Enlistment date: 25 Feb 1814 - Enlistment place: Hallowell - Enlistment period: War - By whom: Fillebrown - Bounty: BLW 22441-160-12 - Land bounty to James Paskiel, brother & other heirs at law of Ezekiel D. Paskiel - On parole at Salem; died at Halifax on 19 Sep 1814.

Paul, David - Private - Company: John Fillebrown - Age: 44 - Height: 5' 6" - Born: New Gloucester, Cumberland, MA - Trade: Farmer - Enlistment date: 15 Jan 1814 - Enlistment place: Hallowell -

Enlistment period: War - By whom: Fillebrown - Bounty: BLW 9021-160-12 - On parole at Salem; sent to Halifax; sent to Fort Wolcott; discharged at New London on 20 Apr 1815.

Porter, Nathan - Private - Company: Jacob Varnum - Age: 33 - Height: 5' 6" - Born: Beverley, MA - Trade: Seaman - Enlistment date: 16 Mar 1814 - Enlistment place: Boston, MA - Enlistment period: War - By whom: Varnum - Bounty: BLW 14725-160-12 - Prisoner of War at Eastport; discharged at Boston on 5 Apr 1815.

Putnam, Perley - Major - Born: MA - Bounty: BLW 12433-160-50 - Commissioned as a major, 40th Infantry, on 19 Jul 1813; on parole of honor at Salem; discharged on 15 Jun 1815.

Ramsdale, Anthony - Private - Company: Jacob Varnum - Age: 43 - Height: 5' 5 3/4" - Born: Salem, Essex County, MA - Trade: Mariner - Enlistment date: 19 Nov 1813 - Enlistment place: Salem, MA - Enlistment period: War - By whom: Lieutenant Manning - Bounty: BLW 14726-160-12 - Prisoner of War at Eastport; discharged on 5 Apr 1815.

Redman, Alonzo - Corporal - 21st Infantry - Company: Marston - Fillebrown - Other regiment: Age: 18 - Height: 5' 6" - Born: Monmouth, MA - Trade: Clothier - Enlistment date: 28 Jun 1812 - Enlistment period: 18 months - Bounty: BLW 12386-160-12 - Discharged on 29 Dec 1813 or 6 Jan 1814; re-enlisted in 40th Infantry at Hallowell by Fillebrown for the war; captured and escaped, at principle rendezvous at Boston; discharged at Boston on 31 Mar 1815.

Richardson, Moses - Private - Company: Jacob Varnum - Age: 24 - Height: 5' 6 1/2" - Born: Cambridge, MA - Trade: Laborer - Enlistment date: 11 Mar 1814 - Enlistment place: Watertown - Enlistment period: War - By whom: Varnum - Bounty: BLW 11583-160-12 - Prisoner of War at Eastport; discharged on 5 Apr 1815.

Rodgers, Elenus - Sergeant - Company: John Fillebrown - Enlistment date: 1 Mar 1814 - Enlistment period: War - Bounty: No bounty found - Captured and escaped, at Houghton.

Simonds, Josiah - Private - Company: John Fillebrown - Age: 25 - Height: 5' 10" - Born: Stowham, Middlesex, MA - Enlistment date: 15 Feb 1814 - Enlistment place: Boston, MA - Enlistment period: War - By whom: 2Lieutenant Blanchard - On parole at Salem; sent to Halifax; sent to Fort Wolcott; Ross Island.

Simpson, Francis - Private - Company: Jacob Varnum - Age: 24 - Height: 5' 8 1/2" - Born: Ware, Hampshire County, MA - Trade: Farmer - Enlistment date: 15 Mar 1814 - Enlistment place: Boston, MA - Enlistment period: War - By whom: Varnum - Bounty: BLW 3854-160-12 - Prisoner of War at Eastport; discharged at Boston on 5 Apr 1815.

Smith, Abraham - Private - Company: John Fillebrown - Age: 21 - Height: 5' 9 1/2" - Born: Waterbrouogh, York, ME - Trade: Farmer - Enlistment date: 22 Feb 1814 - Enlistment place: Hallowell - Enlistment period: War - Pension: SO-445, SC-443 - Bounty: BLW 8474-160-12 - On parole at Machias; deserted from Portland on 21 Jan 1815; discharged at Boston on 13 Mar 1816.

Stiles, Timothy S. - Corporal - Company: Jacob Varnum - Age: 16 - Height: 5' 5 3/4" - Born: Temple, Hillsboro, NH - Trade: Farmer - Enlistment date: 12 Jan 1814 - Enlistment place: Watertown - Enlistment period: War - By whom: Varnum - Bounty: BLW 11684-160-12 - Prisoner of War at Eastport; promoted to sergeant on 29 Mar 1815; discharged at Boston on 5 Apr 1815.

Varnum, Jacob B. - Captain - Company: Jacob Varnum - Pension: SO-17944, SC-5990 - Bounty: BLW

684-160-50 - Commissioned as a captain, 40th Infantry, on 19 Jul 1813; on parole of honor at Boston; discharged on 15 Jun 1815.

Varnum, Samuel M. - Private - Company: Jacob Varnum - Age: 21 - Height: 5' 7" - Born: Draent, Middlesex, MA - Trade: Carpenter - Enlistment date: 17 Dec 1813 - Enlistment place: Concord, NH - Enlistment period: War - By whom: Leonard - Bounty: BLW 9217-160-12 - On parole at Machias; discharged at Portland on 10 Apr 1815.

Wadleigh, Thomas - Sergeant - Company: Jacob Varnum - Age: 36 - Height: 5' 7 1/2" - Born: Sutton, Hillsboro, NH - Trade: Farmer - Enlistment date: 6 Dec 1813 - Enlistment place: Concord, NH - Enlistment period: War - By whom: Leonard - Bounty: BLW 14484-160-12 - Prisoner of War at Eastport; discharged on 5 Apr 1815.

Walker, Seth - Private - Company: John Fillebrown - Age: 19 - Height: 5' 6 3/4" - Born: New Salem, MA - Trade: Cordwainer - Enlistment date: 23 Mar 1814 - Enlistment place: Plymouth - Enlistment period: War - By whom: Lieutenant Pope or Hodge - Pension: Wife Hannah, SO-2063, SC-860, WO-43236, WC-33673 - Bounty: BLW 20008-106-12 - On parole at Salem; sent to Fort Seammel, Portland Harbor; discharged on 10 Apr 1815.

Wilson, Ephraim - Private - Company: Jacob Varnum - Age: 31 - Height: 5' 9" - Born: Mason, Hillsborough, NH - Trade: Blacksmith - Enlistment date: 8 Oct or Nov 1813 - Enlistment place: Northfield - Enlistment period: War - By whom: Capt E Field - Prisoner of War at Eastport, set to Halifax.

Wilson, Levi - Private - Company: John Fillebrown - Age: 37 - Height: 6' - Born: Westmoreland, Cheshire, NH - Trade: Farmer - Enlistment date: 14 Apr 1814 - Enlistment place: Hallowell - Enlistment period: War - Bounty: BLW 19327-160-12 - On parole at Machias; Fort Scammel; discharged on 10 Apr 1815.

Wyman, Amos N. - Musician - Company: Jacob Varnum - Age: 20 - Height: 5' 9 1/2" - Born: Nottingham, Rockingham County, NH - Trade: Blacksmith - Enlistment date: 18 Oct 1813 - Enlistment place: Concord, MA - Enlistment period: War - By whom: Varnum - Bounty: BLW 16346-160-12 - Prisoner of War at Eastport; discharged at New London on 20 Aug 1815.

## Passengers and sailors captured with the 40th Infantry

Adams, Luven
Brown, Thomas
Crenlieu, William
Dill, John
Donnell, John
Dopson, Richard - Pilot
Fasdeck, Peter G. - Mate
Gellmore, John
Graham, Jesse
Greaves, John - Sent to England
Hand, Elijah - Master
Harvey, Samuel
Henson, Samuel
Higgins, Daniel - Passenger
Hill, Levi
Hills, Joseph

Hurst, Charles B. - Passenger
Hussy, Edward - Send to England
Hutchins, Loomis - Sent to England
Lambert, William - Master
Lewis, Henry - Sent to England
Marion, James
Monk, Charles - Sent to England
Monshene, William
Moody, Thomas - Died at Halifax on 23 Feb 1815 from dysentery
Page, Joseph - Mate - Sent to England
Peckham, Polly - Sent to England
Rooney, James - Sent to England
Russell, James - Sent to England
Williamson, Richard - Sent to England
Wood, Edward - Sent to England

## Anatomy of an artillery company:
## Captain Joseph Philips' Company, 2nd Regiment of U.S. Artillery

Tennessee had the privilege of raising an artillery company for the U.S. Army during the War of 1812 as part of the 2nd Regiment of U.S. Artillery. This state is noted for raising the 24th and the 39th Regiments of U.S. Infantry but its artillery company is largely forgotten today.

When William P. Anderson of Tennessee was commissioned a colonel and given command of the 24th Infantry on 6 July 1812, he was also tasked with recruiting an artillery company. Joseph Philips (only one 'L' is his surname) of Rutherford County, Tennessee, was commissioned to be the captain of this company and he began recruiting men in central Tennessee.

Colonel Anderson was ordered on 9 October 1812 to take command of Fort Massac in the Territory of Illinois. The regiment and the artillery company left Nashville shortly after receiving this order, and the units proceeded down the Cumberland River to the Ohio River, probably on flat-bottom boats. Once on the Ohio River, it was a short distance to the fort.

Captain Philips' company consisted of 39 men, far short of the 103 men authorized for an artillery company. Under the Military Act of 26 June 1812, an artillery company was made up of a captain, a first lieutenant, a second lieutenant, four sergeants, four corporals, two musicians and 80 artificers (privates). Another sergeant's position was added to all artillery companies on 3 March 1813.

The position of first lieutenant was never filled in Captain Philip's company. Most likely, the first lieutenant's position was used on the 2nd Artillery's headquarters staff for the adjutant's, the regimental quartermaster's, or the regimental paymaster's position. James Gamble was the company's second lieutenant, and he would perform the duties of a first lieutenant.

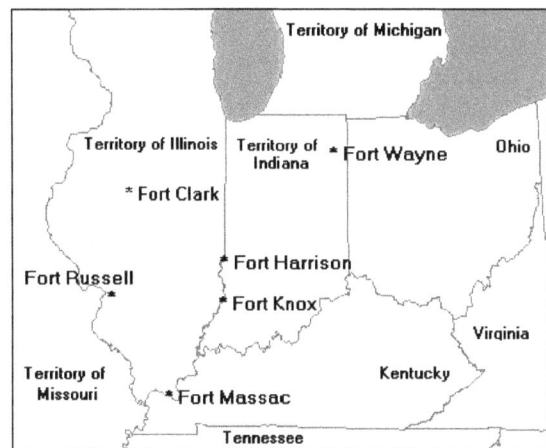

Forts in the old Northwest Territory

The company would be stationed at Fort Massac for a year and a half, and the two officers would continue to recruit artilleryman. Besides recruiting 30 more men from the Fort Massac area, three men were recruited in Missouri, and another 12 men were transferred from the 1st Regiment of Artillery.

The monthly returns of the U.S. Army in the Territory of Indiana during January 1814 showed that Captain Phillips had 70 men in his company at Fort Massac. Although Fort Massac was located in the Territory of Illinois, the fort was the responsibility of the army command at Fort Knox, Vincennes, Indiana..

Captain Philips was given command of Fort Massac on 10 March 1813 when Colonel Anderson received orders to take his regiment to Cleveland, Ohio, in support of Major General William Henry Harrison's Army of the Northwest. The size of the company started to decline after the first of the year as the men from Tennessee who had enlisted for 18-months started to be discharged and they returned home. The company would always be an undersized unit.

Second Lieutenant Joseph Smith of Tennessee was transferred from the 24th Infantry to the 2nd Artillery, and he was assigned to the company on 7 June 1813. He would resigned from the army a few months later on 30 September 1813.

On 30 March 1814, the new military act increased the size of an artillery company to 123 officers and enlisted men. The company was now authorized to have a captain, a first lieutenant, two second lieutenants, an ensign, one quartermaster sergeant, five sergeants, eight corporals, four musicians and 100 privates. Again, the company was not able to recruit to its authorized strength. Third Lieutenant Benjamin Tennile of Kentucky was assigned to the company in May 1814, and he would remain with this unit until the end of the war.

In July 1814, Brigadier General Benjamin Howard ordered Captain Philips to take his company to Fort Clark on the Illinois River in central Illinois. At this time, the company had approximately 45 men, and it was re-enforcing the 34 soldiers already stationed at this fort.

After the war, Captain Philips was not retained by the army and he was discharged. Captain William O. Allen took command of the company before the army reorganized its regiments and companies for peacetime duties. A total of 106 men have been found to have been a assigned to Captain Phillips' company.

According to the comments column in the *Register of Enlistments in the U.S. Army, 1798-1914,* no muster rolls exist for this company prior to 1815. A page-by-page search of this 7,360 document was conducted to recreate a muster roll of Captain Philip's company between 1812 and 1815. Additionally, a page-by-page search of the Military Bounty Land Warrants and the War of 1812 Pension File identified more men not found in the *Register of Enlistments.*

## Re-constructed Muster Roll

Officers
Captain Joseph Philips                                     Second Lieutenant Joseph D. Smith
Second Lieutenant James H. Gamble                  Second Lieutenant Benjamin Tennile

Quartermaster Sergeant
Stewart, Thomas

Sergeants
| Dupriest, Samuel | Hoyt, Fitch | Summers, John |
| Holbrook Sr., John | Storey, John | Symmes, Charles |

Corporals
| Brown, Robert | Hooper, William | Smalley, Josiah |
| Burrow, Jarrel | Parks, Laban | Taylor, Samuel |

Musicians
| Anderson, Samuel | Anderson, William | Hines, William | Tucker, Robert |
| Anderson, Thomas | Duncanson, John | Holbrook Jr., John | |

Privates
| Albright, John | Eveland, Frederick | Lewis, Bazel | Roberts, James |
| Anderson Jr., Joseph | Eveland, Moses | Lloyd, Ephraim | Robertson, Burrell |
| Anderson Sr., Joseph | Furgerson, William B. | Lore, Christopher G. | Robinson, Elisha |
| Baker, William | Grammer, Peterson | Luyster, Garret | Rogers, Edward |
| Ballinger, Moses | Grammer, Pleasant | Mathews, John | Rolston, Samuel |
| Barnett, Jeremiah | Harriott, John | Matson, Thomas | Saunders, George |
| Bird, John | Harris, William | McIntire, John | Scott, Robert |
| Boone, John | Hartgrave, Francis | McKinzie, William | Smith, James |
| Botts, John | Haymaker, John | Mick, Adam | Smith, Samuel B. |
| Bridges, William | Hays, Joseph | Mitchell, James | Sneed, John |
| Brown, Alexander | Hedges, Solomon | Nichols, Ransom | Stratton, Willis |
| Brown, James | Henderson, John | Orr, Alexander | Stigall, Zachariah |
| Burrow, Philip | Herrald, Cader | Parker, Ezekiel | Tolly, James |
| Cagle, Simon | Higgins, Robert | Parnell, Jacob | White, John |
| Clarke, Charles | Hodges, William | Pearson, Abel | White, Willie |

Privates

| | | | |
|---|---|---|---|
| Conley, John R. | Hollyfield, William | Peate, W. F. | White, Stephen |
| Cowen, Andrew | Humphries, Thomas | Ramsey, John | Williamson, John |
| Dailey, Peter | Jones, William | Richards, Joseph | Yocom, Mathias |
| Duke, Abraham | Leatherman, James | Riley, Benjamin | |

## The roster

Albright, John – Private – Age: 26 – Height: 5' 9" – Born: Pennsylvania – Trade: Laborer – Enlisted: 25 Dec 1814 at Fort Clark, IL – Enlistment Period: War - By whom: Joseph Philips – Discharged: 21 Aug 1815 at Fort Clark, IL - Bounty: BLW 831-320-14.

Anderson Jr., Joseph – Private – Age: 15 – Height: 4' 7" – Born: Mecklenburg County, VA – Trade: Laborer – Enlisted: 4 Jan 1813 – Enlistment Period: 5 Years - By whom: Joseph Philips – Discharged: 4 Jan 1818 - Bounty: BLW 20935-320-14.

Anderson Sr., Joseph – Private – Age: 37 – Height: 5' 8 1/2" – Born: Hanover County, VA – Trade: Hatter – Enlisted: 4 Jan 1813 – Enlistment Period: 5 Years - By whom: Joseph Philips – Discharged: 4 Jan 1818 - Bounty: BLW 25384-320-14 – Land bounty to William Anderson & other heirs at law of Joseph Anderson Sr., issued on 10 Jun 1823.

Anderson, Samuel - Fifer - Age: 12 - Height: 4' 3" - Born: Mecklenburg County, VA - Trade: Laborer - Enlisted: 6 Jan 1813 - Enlistment period: 5 Years - By whom: Joseph Philips - Discharged: 6 Jan 1818 - Bounty: BLW 20933-160-12.

Anderson, Thomas - Drummer - Age: 14 - Height: 4' 4 1/2" - Born: Mecklenburg County, VA - Trade: Laborer - Enlisted: 4 Jan 1813 - Enlistment period: 5 Years - By whom: Joseph Philips - Discharged: 4 Jan 1818 – Bounty: BLW 20934-160-12.

Anderson, William - Musician - Age: 10 - Height: 4' 1 1/4" - Born: Halifax County, VA - Enlisted: 3 Jan 1814 at Fort Massac, IL - Enlistment period: 5 Years - By whom: Joseph Philips - Discharged: 3 Jan 1819 at New Orleans, LA – Bounty: BLW 21532-160-12.

Baker, William - Private - Enlisted: 31 Oct 1812 - Enlistment period: 18 Months - Discharged: 30 Apr 1814 – Bounty: BLW 16756-160-50.

Ballinger, Moses - Private - Enlisted: 6 Aug 1812 - Enlistment period: 18 Months - Discharged: 6 Feb 1814 at Nashville, TN.

Barnett, Jeremiah - Private - Enlisted: 2 Aug 1812 - Enlistment period: 18 Months - Discharged: 1 Feb 1814 at Nashville, TN.

Bird, John - Private - Enlisted: 7 Aug 1812 - Enlistment period: 18 Months - Discharged: 7 Feb 1814.

Boone, John - Private - Enlistment date: 22 Sep 1812 - Enlistment period: 18 Months - Discharge date: 22 Mar 1814.

Botts, John - Private - Age: 24 - Height: 5' 7 1/2" - Born: Maryland - Trade: Laborer - Enlisted: 16 Jun 1812 - Enlistment period: 18 Months - By whom: James Gamble - Discharged: 1 May 1819 at Bay of St. Louis, MO – Bounty: BLW 22188-160-12 - Re-enlisted for 5 years at Fort Massac, IL, on 1 May 1814.

Bridges, William - Private - Age: 23 - Height: 5' 8 1/2" - Born: Wake County, NC - Trade: Laborer - Enlisted: 10 Dec 1813 at Fort Massac, IL - Enlistment period: 5 Years - By whom: Joseph Philips - Discharged: 10 Dec 1818 at Bay of St. Louis, MO – Bounty: BLW 21132-160-12.

Brown, Alexander - Private - Enlisted: 10 Jun 1810 - Enlistment period: 5 Years - By whom: Captain Joseph Cross, 1st Artillery - Discharged: 6 Jun 1815 – Transferred from Captain Joseph Cross's Company, 1St Artillery.

Brown, James - Private - Enlisted: 1 Jul 1810 - Enlistment period: 5 Years – Originally enlisted in the Regiment of Artillery and transferred to Captain Philips' Company.

Brown, Robert - Corporal - Age: 24 - Height: 5' 8" - Born: VA - Enlisted: 7 Apr 1810 at Fort Clark, IL - Enlistment period: 5 Years - By whom: Joseph Philips – Originally enlisted in the Regiment of Artillery and transferred to Captain Philips' Company; re-enlisted at Fort Clark, IL, on 5 Mar 1815 for 'During the War' enlistment.

Burrow, Jarrel - Corporal - Enlisted: 25 Jul 1812 - Enlistment period: 18 Months - Discharged: 25 Jan 1814 – Promoted to corporal on 12 Dec 1812.

Burrow, Philip - Private - Enlisted: 23 Jun 1812 - Enlistment period: 18 Months - Discharged: 23 Dec 1813 at Vincennes, IN.

Cagle, Simon - Private - Enlisted: 2 Aug 1812 - Enlistment period: 18 Months - By whom: James Gamble - Discharged: 2 Feb 1814 at Nashville, TN.

Clarke, Charles - Private - Age: 26 - Height: 5' 8 1/2" - Born: Virginia - Enlistment date: 18 Feb 1814 - Enlistment place: Fort Massac, IL - Enlistment period: War - By whom: Joseph Philips - Discharged on 21 Aug 1815 at Fort Clark, IL.

Conley, John R. - Private - Enlisted: 1 Aug 1812 - Enlistment period: 5 Years - Died: 8 May 1814, probably at St. Louis, MO.

Cowen, Andrew - Private - Age: 33 - Height: 5' 8" - Born: Russell County, VA - Enlisted: 6 May 1814 at Fort Massac, IL - Enlistment period: War - By whom: Joseph Philips - Discharged: 21 Aug 1815 at Fort Clark, IL.

Dailey, Peter - Private - Enlistment date: 10 Dec 1809 - Enlistment period: 5 Years - Discharge date: 10 Dec 1814.

Duke, Abraham - Private - Age: 28 - Height: 6' - Born: Rutherford County, NC - Enlisted: 12 Sep 1812 - Enlistment period: 18 Months - Died: 27 Feb 1814 at Fort Massac, IL - Bounty: BLW 26440-160-12 - Re-enlisted 12 Feb 1814 for the war; land bounty to Mary Duke, widow & other heirs at law of Abraham Duke.

Duncanson, John - Fifer - Age: 29 - Height: 5' 9" - Born: Massachusetts - Trade: Soldier - Enlisted: 20 Sep 1814 at Fort Clark, IL - Enlistment period: War - By whom: Joseph Philips - Discharged: 21 Aug 1815 at Fort Clark, IL - Bounty: BLW 15815-160-12.

Dupriest, Samuel - Sergeant - Enlisted: 29 Jun 1812 - Enlistment period: 18 Months - Discharged: 10 Nov 1813 probably at Vincennes, IN.

Eveland, Frederick - Private - Age: 18 - Height: 5' 3" - Born: New York - Trade: Laborer - Enlisted: 14 Apr 1813 at Belle Fontaine, MO - Enlistment period: War - By whom: Captain John Symmes, 1st Infantry - Discharged: 21 Aug 1815 at Fort Clark, IL - Bounty: BLW 15816-160-12.

Eveland, Moses - Private - Age: 41 - Height: 5' 9" - Born: Virginia - Trade: Laborer - Enlisted: 8 Mar 1813 at Belle Fontaine, MO - Enlistment period: War - By whom: Captain John Symmes, 1st Infantry - Discharged: 21 Aug 1815 at Fort Clark, IL - Bounty: BLW 15814-160-12.

Furgerson, William B. - Private - Enlisted: 18 Jun 1812 - Enlistment period: 18 Months - Discharged: 18 Dec 1813 probably at Vincennes, IN.

Gamble, James H. - Second Lieutenant - Commissioning date: 12 Mar 1812 - Discharged: 15 Jun 1815.

Grammer, Peterson - Private - Enlisted: 31 Dec 1813 - Discharged: 28 Apr 1814 - Pension: Wife Martha Whinery, SO-26297, SC-18590; served as a private in Captain Joseph Philips' Company, 2nd Artillery; married 16 Aug 1816 in Bedford County, TN; soldier died on 29 Nov 1885 in Lawrence County, MO; lived in Bedford County, TN, and Lawrence County, MO – Bounty: BLW 1739-160-50.

Grammer, Pleasant - Private - Enlisted: 16 Oct 1812 - Enlistment period: 18 Months - Died: 14 Oct 1813 probably at Vincennes, IN.

Harriott, John - Private - Enlistment date: 19 Jul 1813 - Enlistment period: 5 Years - By whom: Lieutenant Symmes - Died: 10 Jul 1817 in hospital at New Orleans, LA.

Harris, William - Private - Enlisted: 24 Jul 1812 - Enlistment period: 18 Months - Discharged: 24 Jan 1814.

Hartgrave, Francis - Private - Enlistment date: 3 Sep 1812 - Enlistment period: 18 months - Died: 17 Oct 1813.

Haymaker, John - Private - Enlisted: 9 Sep 1809 - Enlistment period: 5 Years - Discharged: 9 Sep 1814 – Transferred from Captain Joseph Cross's Company, 1St Artillery, June 1813 – BLW 26513-160-50 Cancelled.

Hays, Joseph - Private - Enlisted: 12 Sep 1812 - Enlistment period: 5 Years - Died: 9 Jun 1815 probably at Fort Clark, IL.

Hedges, Solomon - Private - Age: 28 - Height: 5 10 - Born: Virginia - Enlistment date: 1 Apr 1810 - Enlistment place: Fort Clark, IL - Enlistment period: 5 Years - By whom: Joseph Philips - Re-enlisted 28 Feb 1815 for the war.

Henderson, John - Private - Age: 28 - Height: 5' 9" - Born: Granville County, NC - Trade: Farmer - Enlisted: 1 Aug 1812 - Enlistment period: 5 Years - By whom: Lieutenant Taliaferro Richards, 24th Infantry - Discharged: 10 Aug 1817 at New Orleans, LA – Bounty: BLW 13150-160-12.

Herrald, Cader - Private - Enlisted: 14 Aug 1812 - Enlistment period: 18 Months - Discharged: 14 Feb 1814.

Higgins, Robert - Private - Age: 43 - Height: 6' 1/2" - Born: Caroline County, VA - Enlisted: 13 Feb 1814 - Enlistment period: War - Discharged: 21 Aug 1815 at Fort Clark, IL - Bounty: BLW 25028-160-12.

Hines, William – Fifer – Age 15 – Height: 4' 8" – Born: Virginia – Trade: Farmer – Enlisted: 20 Dec 1814 at Fort Clark, IL – Enlistment Period: War – By Whom: Joseph Philips – Discharged: 21 Aug 1815 at Fort Clark, IL - Bounty: BLW 632-320-14.

Hodges, William - Private - Age: 23 - Height: 6' 1" - Born: Surrey County, NC - Enlistment date: 17 Jan 1814 - Enlistment place: Fort Massac, IL - Enlistment period: War - By whom: Joseph Massac - Discharged: 21 Aug 1815.

Holbrook Jr., John - Drummer - Age: 13 - Height: 4' 7" - Born: Boston, MA - Trade: Laborer - Enlistment date: 21 Sep 1821 - Enlistment period: 5 Years - By whom: Lieutenant Allen - Discharged at Pettite Coquille.

Holbrook Sr., John - Sergeant - Age: 38 - Height: 5' 9" - Born: Boston, MA - Trade: Laborer - Enlistment date: 21 Sep 1812 - Enlistment period: 5 Years - By whom: Lieutenant Allen - Discharged: 21 Sep 1817 at Pettite Coquille.

Hollyfield, William - Private – Enlisted: 30 Jan 1814 – Enlistment period: 5 Years – By whom: Joseph Philips – Deserted on 16 Aug 1819 at Fort Charles, LA - Pension: SO-23706, SC-17246; served as a private in Captain Joseph Phillips' Company, US Artillery, and as a corporal in Captain William O. Allens' Company, Corps of Artillery.

Hooper, William - Corporal - Enlisted: 19 Jul 1812 - Enlistment period: 18 Months - By whom: James Gamble - Discharged: 19 Jan 1814 at Nashville, TN.

Hoyt, Fitch - Sergeant - Enlistment date: 21 Aug 1812 - Enlistment period: 18 months - Discharge date: 21 Feb 1814.

Humphries, Thomas - Private - Age: 16 - Height: 5' 5 1/2" - Born: Nelson County, KY - Trade: Laborer - Enlistment date: 28 Jul 1813 - Enlistment place: Fort Massac, IL - Enlistment period: 5 Years - By whom: Joseph Philips - Discharged: 23 Jul 1818 at Pettite Coquille.

Jones, William - Private - Enlisted: 16 Oct 1812 - Enlistment period: 18 Months - Discharged: 16 Apr 1814.

Leatherman, James - Private - Age: 16 - Height: 5' - Born: Kentucky - Trade: Laborer - Enlisted: 10 May 1814 at Fort Massac, IL - Enlistment period: War - By whom: Joseph Philips - Discharged: 10 May 1819 from Bay of St. Louis, MO – Bounty: BLW 22338-160-12.

Lewis, Bazel - Private - Age: 44 - Height: 5' 6" - Born: Maryland - Enlisted: 11 Feb 1814 at Fort Massac, IL - Enlistment period: War - By whom: Joseph Philips - Discharged: 21 Aug 1815 at Fort Clark, IL - Bounty: BLW 25073-160-12.

Lloyd, Ephraim - Private - Enlisted: 25 Jul 1812 - Enlistment period: 18 Months - Discharged: 25 Jan 1814.

Lore, Christopher G. - Private - Enlisted: 1 Oct 1813 - Enlistment period: 18 Months - Discharged: 1 Apr 1814.

Luyster, Garret - Private - Enlisted: 28 Oct 1813 - Enlistment period: 5 Years.

Mathews, John - Private - Age: 28 - Height: 5' 9" - Born: Orange County, NY - Trade: Shoemaker - Enlisted: 6 Mar 1810 - Enlistment period: 5 Years - Discharged: 21 Aug 1815 at Fort Clark, IL.

Matson, Thomas - Private - Enlisted: 23 Sep 1813 - Enlistment period: 18 Months - Discharged: 26 Mar 1814.

McIntire, John - Private - Age: 49 - Height: 6' - Born: Scotland - Enlisted: 14 Mar 1814 at Fort Massac, IL - Enlistment period: War - By whom: Joseph Philips.

McKinzie, William - Private - Age: 21 - Height: 5' 9" - Born: Madison County, KY - Enlisted: 3 Dec 1813 at Fort Massac, IL - Enlistment period: 5 Years - By whom: Joseph Philips - Discharged: 3 Dec 1818 – Bounty: BLW 24273-160-12.

Mick, Adam - Private - Age: 26 - Height: 5' 7" - Born: Virginia - Enlisted: 19 Dec 1814 at Fort Clark, IL - Enlistment period: War - By whom: Joseph Philips - Discharged: 21 Aug 1815 at Fort Clark, IL - Bounty: BLW 26387-160-12 Cancelled (Entitled to 320 acres); BLW 1060-320-14.

Mitchell, James - Age: 32 - Height: 6' - Born: Brunswick County, VA - Enlisted: 27 Jan 1814 at Fort Massac, IL - Enlistment period: War - By whom: Joseph Philips

Nichols, Ransom - Private - Age: 21 - Height: 5' 9" - Enlisted: 30 Oct 1812 - Enlistment period: 18 Months - Discharged: 30 Apr 1814 - Re-enlisted at Fort Massac, IL, 3 May 1814 for during the war service – Bounty: BLW 23097-160-50.

Orr, Alexander - Private - Enlisted: 16 May 1812 - Enlistment period: 5 Years - By whom: James Gamble – Deserted and discharged: 9 Jan 1813 - Bounty: BLW 27637-160-42.

Parker, Ezekiel - Private - Enlisted: 24 Sep 1811 - Enlistment period: 5 Years - By whom: Captain Joseph Cross, 1$^{st}$ Artillery - Discharged: 31 Dec 1816 – Transferred from Captain Joseph Cross's Company, 1$^{St}$ Artillery.

Parks, Laban - Corporal - Enlisted: 5 Jul 1812 - Enlistment period: 5 Years - By whom: James Gamble - Discharged: 5 Jul 1817 – Bounty: BLW 22348-160-12.

Parnell, Jacob - Private - Age: 21 - Height: 6' - Born: Abberville County, SC - Enlisted: 3 Mar 1814 at Fort Massac, IL - Enlistment period: War - By whom: Joseph Philips – Bounty: BLW 20245-16-12.

Pearson, Abel - Artificer - Enlisted: 21 Sep 1812 - Enlistment period: 18 Months - Discharged: 2 Mar 1814.

Peate, W. F. - Private - Age: 26 - Height: 5' 5 1/2" - Born: England - Trade: Laborer - Enlisted: 1 Oct 1812 - Enlistment period: 5 Years - By whom: Joseph Philips - Discharged: 1 Oct 1817 at Petite Coquille, LA.

Peeler, Pleasant - Private - Enlistment date: 11 Aug 1812 - Enlistment period: 18 Months - Discharged: 11 Feb 1814.

Phelps, William - Artificer - Enlisted: 16 Jun 1812 - Enlistment period: 18 Months - Discharged: 16 Dec 1813.

Philips, Joseph - Captain - Commissioning date: 12 Mar 1812 - Discharged: 15 Jun 1815 - Pension: Wife

Dorothy, WO-19029, WC-13053; served as a captain in the 2nd Artillery.

Philips, Thomas - Private - Enlisted: 8 Jun 1812 - Enlistment period: 5 Years - Died: 14 Feb 1813.

Pixley, John - Private - Age: 22 - Height: 5' 10" - Born: Pennsylvania - Trade: Laborer - Enlisted: 7 Mar 1813 at Belle Fontaine, MO - Enlistment period: War - Discharged: 21 Aug 1815 at Fort Clark, IL - Bounty: BLW 15813-16012.

Price, William - Private - Enlisted: 4 Jun 1812 - Enlistment period: 18 Months - Discharged: 4 Dec 1813 probably at Vincennes, IN.

Radford, Elijah - Private - Enlisted: 18 Jun 1812 - Enlistment period: 18 Months - Discharged: 10 Jan 1814, probably at Vincennes, IN.

Ramsey, Daniel - Private - Enlisted: 5 Oct 1812 - Enlistment period: 18 Months - By whom: James Gamble – Discharged: Apr 1814.

Ramsey, John - Private - Age: 33 - Height: 5' 11 1/2" - Born: Overton County, VA - Enlisted: 7 May 1814 at Fort Massac, IL - Enlistment period: War - By whom: Joseph Philips - Discharged: 21 Aug 1815 at Fort Clark, IL.

Rhea, Matthew - Private - Age: 39 - Height: 5' 6 1/2" - Born: Montgomery County, PA - Enlisted: 25 Aug 1813 at Fort Massac, IL - Enlistment period: 5 Years - By whom: Joseph Philips – Discharged 25 Aug 1818 at Petite Coquille, LA – Bounty: BLW 20932-160-12.

Richards, Joseph - Private - Enlisted: 21 Jan 1812 - Enlistment period: 5 Years - By whom: Captain Joseph Cross, 1st Artillery – Transferred from Captain Joseph Cross's Company, 1st Artillery.

Riley, Benjamin - Private - Enlisted: 13 Sep 1809 - Enlistment period: 5 Years - Discharged: 13 Sep 1814 probably at St. Louis, MO.

Roberts, James - Private - Enlisted: 29 Aug 1812 - Enlistment period: 18 Months - Died: May 1813 at Fort Massac, IL.

Robertson, Burrell - Private - Age: 23 - Height: 5' 6 1/2" - Born: Orange County, NC - Enlistment date: 14 Dec 1813 - Enlistment place: Fort Massac, IL - Enlistment period: 5 Years - By whom: Joseph Philips - Discharge date: 14 Dec 1818 at Bay of St. Louis, LA.

Robinson, Elisha - Private - Enlisted: 9 Jun 1812 - Enlistment period: 18 Months - Discharged: 9 Dec 1813 at Fort Massac, IL.

Rogers, Edward - Private - Enlisted: 1 Feb 1810 - Enlistment period: 5 Years - Discharged: 1 Feb 1815.

Rolston, Samuel - Private - Age: 28 - Height: 6' 3 1/2" - Born: Cumberland County, PA - Enlisted: 1 Mar 1814 at Fort Massac, IL - Enlistment period: War - By whom: Joseph Philips - Discharged: 21 Aug 1815 - Bounty: BLW 25498-160-12.

Saunders, George - Private - Enlisted: 7 Sep 1812 - Enlistment period: 18 Months - Discharged: 7 Mar 1814.

Scott, Robert - Private - Enlisted: 7 Nov 1809 - Enlistment period: 5 Years - Discharged: 5 Nov 1814 probably at St. Louis, MO.

Smalley, Josiah - Corporal - Enlisted: 20 Dec 1809 - Enlistment period: 5 Years - By whom: First Lieutenant Hezekiah Johnson, 1st Infantry - Discharged: 20 Dec 1814 probably at St. Louis, MO.

Smith, James - Private - Enlisted: 30 Jul 1812 - Enlistment period: 18 Months - Discharged: 30 Jan 1814.

Smith, Joseph D. - Second Lieutenant - Commissioning date: 12 Mar 1812 – Resigned on 30 Sep 1813 - Transferred from 24th Infantry on 7 Jun 1813.

Smith, Samuel B. - Private - Enlistment date: 7 Jun 18111 - Enlistment period: 5 Yrs - By whom: Joseph Cross - Transferred from Captain Joseph Cross's Company, 1st Artillery.

Sneed, John - Private - Enlisted: 9 Oct 1812 - Enlistment period: 5 Years - By whom: James Gamble - Discharged: 9 Oct 1817 at Petite Coquille, LA – Bounty: BLW 22349-160-12 (Arkansas) 14 Aug 1819.

Stewart, Thomas - Quartermaster Sergeant - Enlistment date: 27 Jun 1811 - Enlistment period: 5 Years - By whom: Joseph Cross - Transferred from Captain Cross's Company, 1st Artillery; discharged and re-enlisted.

Stratton, Willis - Private - Age: 21 - Height: 5' 10" - Born: Amherst County, VA - Enlisted: 6 Dec 1813 at Fort Massac, IL - Enlistment period: 5 Years - By whom: Joseph Philips - Deserted at Bay of St. Louis, MO, on 4 Oct 1817.

Stigall, Zachariah - Private - Age: 23 - Height: 5' 7 1/2" - Born: Halifax County, VA - Trade: Laborer - Enlisted: 6 Dec 1813 at Fort Massac, IL - Enlistment period: 5 Years - By whom: Joseph Philips - Discharged: 6 Dec 1818 at Ripley Barracks, Bay of St. Louis, MO - Bounty: BLW 21537-160-12.

Storey, John - Sergeant - Enlistment date: 24 Aug 1809 - Enlistment period: 5 Yrs - Discharged: 24 Aug 1814 - Transferred from Captain Joseph Cross's Company, 1st Artillery, on 30 Jun 1812; discharged as a private.

Summers, John - Sergeant - Enlisted: 24 Dec 1811 - Enlistment period: 5 Years – Discharged: probably at New Orleans, LA.

Symmes, Charles - Sergeant - Age: 37 - Height: 5' 9" - Born: Loudon County, VA - Trade: Soldier - Enlisted: 19 Apr 1813 - Enlistment period: War - Discharged: 21 Aug 1815 at Fort Clark, IL – Bounty: BLW 20815-160-12.

Taylor, Samuel - Corporal - Enlistment date: 23 Sep 1809 - Enlistment period: 5 Years - Discharged: 23 Sep 1814 - Transferred from Captain Joseph Cross's Company, 1st Artillery.

Tennile, Benjamin - Second Lieutenant - Enlisted: 23 May 1814 as Third Lieutenant - Promoted: Second Lieutenant 31 Aug 1814 – Discharged: 15 Jun 1815.

Tolly, James - Private - Enlisted: 7 Oct 1812 - Enlistment period: 18 Months - Discharged: 7 Apr 1814.

Tucker, Robert - Drummer - Age: 21 - Height: 6' 1" - Born: Virginia - Enlisted: 20 Sep 1814 at Fort Clark, IL - Enlistment period: War - By whom: Joseph Philips. – Pension: SO-27182, SC-20707, drummer in Captain Joseph Cross' Company, 1st Artillery.

White, John - Private - Age: 24 - Height: 5' 10" - Born: Warren County, NC - Enlisted: 6 May 1814 at Fort Massac, IL - Enlistment period: War - By whom: Joseph Philips - Discharged: 13 Jul 1815 - Bounty: BLW 27730-160-42 Cancelled; BLW 27938-160-42.

White, Willie - Private - Enlisted: 13 Jun 1812 - Enlistment period: 18 Months - Discharged: 13 Dec 1813 at Fort Massac, IL.

White, Stephen – Artificer – Enlisted: 15 Jul 1812 – Enlistment period: 18 Months – Discharged: 15 Jan 1814.

Williamson, John - Private - Enlistment date: 8 Jan 1812 - Enlistment period: 5 Years - By whom: Joseph Cross - Transferred from Captain Joseph Cross's Company, 1st Artillery.

Yocom, Mathias - Private – Enlisted: 19 Jun 1812 – Enlistment period: 5 Years - Discharged 19 Jun 1817 at Petite Coquille, LA - Pension: SO-18849, SC-12689; served in Captains Joseph Phillips' and W. Allen's Companies, U.S. Artillery – Bounty: BLW 24465-160-12.

## Bibliography

*Transactions of the Illinois State Historical Society for the Year 1903*, (Phillips Bros., State Printers: Springfield, IL 1094), <u>Old Fort Massac</u>, pp. 61-62.

Heitman, Francis B., *Historical Register and Dictionary of the United States Army From Its Organization, September 29, 1789, to March 2, 1903*, Volumes I and II, (Genealogical Publishing Company, Baltimore, Maryland: 1994).

Powell, Colonel William H. (US Army), *List of Officers of the Army of the United States from 1779-1900*, (L. R. Hamersly & Co.: New York 1900).

*Register of Enlistments in the U.S. Army, 1798-1914*; Records of the Adjutant General's Office, 1780's-1917, Record Group 94; National Archives, Washington, D.C.

United States. Bureau of Land Management, General Land Office Records. Automated Records Project; *Federal Land Patents*, State Volumes. http://www.glorecords.blm.gov/. Springfield, Virginia: Bureau of Land Management, Eastern States, 2007.

*War of 1812 Military Bounty Land Warrants, 1815-1858*; National Archives Microfilm Publication M848; Records of the Bureau of Land Management, Record Group 49; National Archives, Washington, D.C.

*War of 1812 Pension Applications*; National Archives Microfilm Publication M313; Records of the Department of Veterans Affairs, Record Group Number 15; National Archives, Washington D.C.

Carter, Clarence Edwin, *The Territorial Papers of the United States*, Volume XVI, The Territory of Illinois 1809-1814, (Government Printing Office: Washington, DC 1948), pp 444-446, <u>Letter from Brigadier General Benjamin Howard to Secretary of War John Armstrong</u>, 15 Jul 1814.

*Duncan McArthur Papers*, Library of Congress, 1922, Microfilm 47, reel 2, volumes 4-5, 22 September 1813 – 4 March 1814, Monthly Returns for the Territory of Indiana, 31 January 1814, Ohio Historical Society, Columbus, Ohio.

## The Baltimore Volunteers:
Maryland's elite militia company

Few militia companies can hold a candle to the feats and accomplishments achieved by the men who served in the Baltimore Volunteers, a volunteer militia company raised in Baltimore, Maryland, during the War of 1812.

Most militia companies that were called to duty by the federal government served as construction troops, performed guard duties in the many forts and batteries along our nation's shorelines, or were used to guard prisoners, transport supplies and deliver dispatches for the regular army. Other militia companies were called to duty as combat troops but when the time came to attack the enemy in their own domain, they refused to cross the borders of this country. On the other hand, of the few militia companies that did serve shoulder-to-shoulder with the U.S. Army only a handful of these companies had such a colorful and distinctive career as the Baltimore Volunteers.

The Baltimore Volunteers was formed during the summer of 1812 to serve with the U.S. Army as part of the U.S. Voluntary Corps, which was created under the Act of 6 February 1812. The Corps was created to supplement the army until all of the army regiments were at full strength. This Corps was made up of militiamen who would serve a one-year enlistment with the army and they were not under the control of any state or territorial government.

All members of the Corps had to provide their own uniforms while the federal government would arm and equipped these volunteers. The soldiers were covered by the same rules and regulations as the regular army and these men were entitled to the same benefits as regular soldiers except for receiving land bounties and the yearly clothing issuances.

Initially, most of the volunteer companies were attached to individual army regiments. Many of these companies would serve as light infantry or as rifle units within the various army infantry regiments that were created during 1812. Other independent companies would be organized as volunteer regiments and provide the same services to army brigades.

The Baltimore Volunteers was organized under the command of Captain Stephen H. Moore in August 1812 as a light infantry company made up of 111 men. Once the Baltimore Volunteers was accepted into federal service in early September 1812 they were ordered to report to the Niagara Theater in New York to begin operations with the Army of the North. The company was ordered to proceed to Carlisle, Pennsylvania, where other volunteer companies from Pennsylvania would join them.

Captain Moore's company arrived at Carlisle on Saturday, 2 October 1813. This site was chosen as the rendezvous point for the volunteers from Baltimore, Philadelphia and Carlisle. Located at Carlisle was the U.S. Army's Carlisle Barracks where the volunteers were armed and equipped. The Carlisle Barracks is the second oldest active U.S. Army post. The British had established this post during the French and Indian War and the Continental Army used it during the American Revolution. The post has been active in the current U.S. Army since 1801.

The volunteers arrived in Buffalo, New York, on Sunday, 8 November 1812, after their march from Carlisle. On November 9th, Ensign Thomas Warner of the *Baltimore Volunteers* went down to Black Rock, New York, along the Niagara River to visit Captain Nathan Towson of the 2nd Regiment of U.S. Artillery. Towson was a native of Maryland and by all indications a friend of Ensign Warner's before the war. He also saw Second Lieutenant Joseph Hook of the same regiment who was another native of Maryland.

Ensign Warner also paid his respects to Colonel William Henry Winder, commander of the 14th Regiment of U.S. Infantry. This regiment was headquartered in Baltimore, Maryland, and had been assigned to the Buffalo area. Winder was the commander of this regiment from 6 July 1812 to 12 March 1813 before he was promoted to brigadier general and went on to greater fame during the war.

The *Baltimore Volunteers* were assigned to the battalion commanded by Major Darby Noon of Lieutenant Colonel Francis McClure's regiment of the U.S. Voluntary Corps.

Some of the volunteers from Baltimore, led by Second Lieutenant Baptist Irvine, who was the editor of the newspaper, the Baltimore Whig, got into a dispute with a hotel owner in Buffalo over politics. This started the famed Buffalo riots on the evening of 25 November 1812. Some of the men from the Baltimore Volunteers tried to burn down the hotel. Three men in the mob that was formed during this incident were killed. None of these men were from Maryland. The army and several other volunteer companies had to put down the riot.

The Baltimore Volunteers' first engagement with the enemy occurred during two attempted invasions of Upper Canada staged from Buffalo on 28 November 1812 and again on 1 December 1812. The company was a part of a 6-company regiment of volunteers under the command of Colonel McClure. In Lossing's Pictorial Field-Book of the War of 1812 he said, "On the 27th of November, when [Brigadier General Alexander] Smyth called the troops to a general rendezvous at Black Rock, they numbered about four thousand five hundred. They were composed of his own regulars, and the Baltimore Volunteers under Colonel Winder, the Pennsylvania Volunteers under General Tannehill, and the New York Volunteers under General Peter B. Porter."

Only part of the invasion force of 3,000 regular soldiers and volunteers made it across the Niagara River on 28 November 1812 before the British discovered the invasion and counter-attacked. After a British artillery battery was put out of commission the Americans were forced to withdraw back across the river.

The invasion, called the Battle of Frenchman's Creek, occurred between today's City of Niagara Falls and Fort Erie in Ontario. The purpose of the invasion was to secure the upper Niagara River area for the United States. The U.S. Army had previously been defeated during the Battle of Queenston Heights on 13 October 1812 in another failed invasion in the lower part of the river.

The Secretary of War John Armstrong wrote a letter on 10 February 1813 to the commander of the Army of the North, Major General Henry Dearborn, outlining the spring campaign for the main invasion of Canada. The army was being staged at Sackets Harbor, New York, and it would attack Kingston, Upper Canada, and then proceed to York, Upper Canada. Once these two objectives had been taken then the army would attack Forts George and Erie in Upper Canada.

Besides the two regular army brigades under the commands of Brigadier Generals Joseph Bloomfield and John Chandler, there would be five detachments of U.S. Volunteers from Philadelphia, Baltimore, Carlisle, Greenbush (New York) and Sackets Harbor. The volunteers numbered a total of 1,525 men: Philadelphia, 400 men; Baltimore, 300 men; Carlisle, 200 men; Greenbush, 400 men; and Sachets Harbor, 250 men. It is not known how accurate these figures are since both the army and the Corps had company structures of 109 men. It appears that there were at least three companies from Baltimore that served in the U.S. Voluntary Corps in New York. The names of these other two companies and their commanders are not known at this time.

The *Baltimore Volunteers* were stationed on the Niagara frontier during the winter of 1812-1813. Four men from this company died from disease or illness and they were buried in the Buffalo, New York. They were Samuel Bowstradst, James Auld, John Hanna and John R. Briers, all privates. These men were probably buried along with approximately 300 other men who had died from camp diseases during this winter in a cemetery that is now a meadow in the Delaware Park in Buffalo. There is a monument in this park dedicated to their memory.

There were a number of army and navy facilities built in and around Buffalo during the War of 1812. In what is now downtown Buffalo was Fort Tompkins and the Flint Hill Camp. At Flint Hill Camp were an encampment and an army hospital. The *Baltimore Volunteers* were probably billeted at this facility. In the area were a number of artillery batteries and a blockhouse protecting the naval yard at Black Rock. To the east of Buffalo were the Cheektowaga Barracks and the Tonawanda Blockhouse. At Cheektowaga was a militia barracks which was later used as an army hospital. There is also a War of 1812 cemetery located at this facility.

The Baltimore Volunteers received their marching orders for central New York in late January 1813. An article in the Buffalo Gazette states, "On Wednesday last orders came on this place to march the U.S. Volunteers, under Capt. Moore and Lieuts. Doyle and Marshall, to Utica. Arrangements were accordingly made to march the Pennsylvania Volunteers, Lieut. Marshall; and Albany Greens, Lieut. Doyle, on Sunday last, the Baltimore Volunteers on Monday. But in consequence of the flag on Saturday [giving an account of General Winchester's Defeat at the River Raisin] the orders were countermanded. The men are again ordered to march, tomorrow morning."

Major General Dearborn wrote to the Secretary of War on 5 April 1813 stating that the troops from Maryland and Philadelphia had arrived at Albany, New York, the day before and that they would move westward in a day or two. The *Baltimore Volunteers* actually arrived at Greenbush, which was across the Hudson River from Albany. The headquarters for the Army of the North was located at a post called the Greenbush Cantonment. This facility contained barracks, officer's quarters, hospital, commissary, arsenal, armory, guardhouses, magazine, stables and a parade ground.

The *Baltimore Volunteers* arrived at Sackets Harbor on Saturday, April 17, 1813, after marching from Greenbush on a route that had been covered by ice and snow. Colonel McClure's regiment was divided and Captain Moore's company plus the *Albany Greens* were attached to Brigadier General Pike's brigade of the regular army. By now the company had 65 effective men meaning that the other 46 men were either on detached duty, were sick or had died.

The *Albany Greens* was probably a rifle company. Many of the volunteer companies had 'blue' or 'green' has part of their names. This signified the color of the uniforms: blue for infantry and green for rifle. It is interesting to note that the British called the *Baltimore Volunteers* the "Baltimore Bloodhounds." This may have indicated that the company did participate in the invasion staged from Buffalo and that the company did fight the British without taking any casualties. It appears that the *Baltimore Volunteers* had earned the respect of the British.

The assault on Kingston, Upper Canada, had been abandoned and the American army instead began to gather troops and supplies for an attack on York, Upper Canada. On 27 April 1813 the American army landed west of York. Major Benjamin Forsyth and his company of riflemen from the Regiment of U.S. Rifles were the first to land and they immediately came under heavy fire from the defending British forces. The volunteers landed in the first boats of the main body of troops and the volunteers headed towards the woods to out flank the enemy. The volunteers, including the *Baltimore Volunteers*, paralleled the army from the woods as the army approached York and attacked the local fort.

The Americans captured the fort, the naval facilities, and the town of York. Two government buildings and a couple of homes were set afire by looters. The army did not have orders to destroy the village. The army left York, Upper Canada, on 8 May 1813 and arrived off Four Mile Creek, New York, just east of Fort Niagara.

Captain Moore lost a leg during the battle while Second Lieutenant Baptist Irvine and Private Thomas Hazeltine were badly wounded. Hazeltine would later die from his wounds nine or ten days after the battle. He is probably buried at the military cemetery in Sackets Harbor (if one of the naval ships had been used as a hospital ship and had carried the wounded back to New York). Also wounded from the company was Private Edward Edwards.

On 27 May 1813 the American army left Four Mile Creek with fresh troops and supplies on board the U.S. naval squadron. The army landed west of Fort George, Upper Canada, and the *Baltimore Volunteers* and the *Albany Volunteers* were used to flank the enemy lines. The Americans overwhelmed the British defenders and the British were forced to abandon Fort George and the village of Newark. Word was sent to the other British facilities along the Niagara River and these facilities were also abandoned, including Fort Erie. The British forces retreated to the western end of Lake Ontario.

Sergeant Robert Kent, Private Frederick McCombs and Private William Porter received wounds during this battle. The wounded men were probably sent across the Niagara River to Fort Niagara in New York, which had a post hospital.

After Fort George fell, the *Baltimore Volunteers* were a part of the U.S. forces that captured Queenston Heights. The volunteer regiment was then used to guard the wounded on the main battlefield. They probably also helped bury the dead from both sides. The *Baltimore Volunteers* returned to Four Mile Creek, where they remained for next three or four weeks. Later, the *Baltimore Volunteers* were stationed at Lewiston and Schlosser, New York, along the Niagara River on guard duty. The company probably remained at these two villages until they were discharged from the army.

The company was disbanded on 7 September 1813, one year after being accepted into federal service, and the men headed home. In one short year the *Baltimore Volunteers* were involved in one riot and three invasions of Upper Canada (now Ontario). One man had died from wounds received during battle while another seven men were wounded. Five men had died of illnesses and diseases while serving in New York.

The *Baltimore Volunteers* served during a time when most militia companies proved to be unreliable. The company was a match to any regular army unit and they served proudly during a critical part of this war.

## Captain Stephen H. Moore's Company, U.S. Volunteers

Ahrens, John D. - Private
Andrews, Nathaniel - Private
Armitage, Joseph - Private - Deserted on 13 Dec 1812
Armstrong, David - Private
Auld, James - Sergeant - Died on 3 Jan 1813
Baker, Christian - Private
Baker, Henry - Private
Barr, Daniel - Private - Deserted on 10 Dec 1812
Bennett, John B. - Private - Deserted on 5 Dec 1812
Bishop, Henry - Private - Land bounty to James Legg, John Henry Legg, William E. Legg, and Charles M. Legg, the nephews and only heirs at law of Henry Bishop - Bounty: BLW 27388-160-12 - Died on 6 Aug 1813
Bowers, Daniel - Private
Bowstred, Samuel - Private - Died at Buffalo on 6 Dec 1812
Boyle, Edward J. - Private
Briers, John R. - Private - Died at Buffalo on 15 Jan 1813
Caffry, John R. - Private
Camper, Jonathan - Private
Chapman, Jonathan - Private
Coates, Daniel - Private - Discharged on 10 Dec 1812, unfit for service
Cochran, James J. L. M. - Private
Cooke, Levin - Private
Craig, George - Corporal
Crane, Joseph L. - Corporal
Daley, John - Private
Dawes, James - Private
Day, Cornelius - Private
Delahay, Henry - Private
Deppisen, John C. - Private
Dixon, John - Private
Doughty, Joseph - Private
Edwards, Aquila - Corporal
Edwards, Edward - Private - Land bounty to Aquila Edwards & other heirs at law of Edward Edwards - Bounty: BLW 26987-160-12 - Killed at York, UC, on 27 Apr 1813
Eichelberger, Peter - Private
Elliott, Benjamin - Private
Evans, George - Sergeant
Fife, Andrew H. - Private
Fitch, Henry - Private

Fonder, Joseph - Private
Forman, Daniel - Private
Foy, Gregory - Sergeant
Frederick McCombs - Private
Gamble, Nicholas - Private
Gardiner, Peter - Private
Gardiner, Samuel - Private
Garraway, James - Private
George, Ezekiel - Private
Gill, John P. - Private - Discharged as unfit for service
Gongers, Peter C. - Private
Grayson, John - Ensign - Bounty: BLW 5138-160-50 - Promoted from private to ensign on 1 Mar 1813
Hanna, John - Private - Died at Buffalo on 13 Jan 1813
Hayes, George - Private
Hayes, Nicholas - Private
Hayley, Edward - Private
Hazletine, Thomas - Private - Wounded at York, UC, on 27 Apr 1813
Hilton, James - Private
Howard, John - Corporal
Irvine, Baptist - Second Lieutenant - Wounded at York, UC, on 27 Apr 1813
Jarvis, William H. - Corporal
Jones, Thomas - Private
Keeves, William - Private
Keller, Conrad - Private
Kelley, Patrick - Private
Kelnor, George - Private
Kent, Robert - Sergeant - Wounded at Newark, UC, on 27 May 1813
Kessner, John - Private
Kleinhans, George H. - Private
Knott, George - Private
Kohlstradst, Benjamin - Private
Laurence, Joseph - Private
Maxwell, John Y. - Private
Maxwell, William - Private
McAllister, Robert - Corporal
McCracken, John - Private
Melcher, Isaac - Private
Merchant, Richard - Private
Moore, Stephen H. - Captain - Land bounty to Jane Moore, widow of Stephen H. Moore - Bounty: BLW 9756-160-50 - Wounded at York, UC, on 27 Apr 1813
Myers, John - Private
Newburgh, John V. - Private - Deserted on 5 Dec 1812
Patterson, John - Private
Penman, John - Private
Peregoy, William - Private
Peters, William - Private
Piercal, Henry - Private
Pike, John H. - Private
Pocock, Thomas - Private
Porter, William - Private - Wounded at York, UC, on 27 Apr 1813
Price, Thomas - Private

Randall, William - Private - Placed on the pension rolls on 7 Sep 1820
Rattle, John - Private
Ryne, William - Private
Sadler Jr., Joseph - Private
Selby, Micajah - Private
Sherman, Lewis - Musician
Sinard, John - Private
Sinton, Francis - Private
Small, John D. - Private
Small, John J. - Private
Speakes, Edward L. - Private
Stoutsberger, Andrew - Private
Tracey, William - Private - Discharged at Baltimore on 24 Sep 1812, unfit for service
Tufts, Samuel - Private
Underwood, James - Private
Van Burgen, William D. - Private
Warner, Thomas - Ensign - Land bounty to Mary Ann Warner, widow of Thomas Warner - Bounty: BLW 7691-160-50
Watkins, William - Private
Webb, William - Private - Discharged on 31 May 1813, inability
Welch, Moses - Private
Williams, John - Private
Williams, Thomas L. – Private

# Bibliography

*American Forts, Eastern United States and Territories*, http://www.geocities.com/naforts/pasocentral

*American State Papers*, Documents, Legislative and Executive of the Congress of the United States, volume 1, military affairs, (Washington, D.C.: Gales and Seaton, 1832).

Cruikshank, Ernest, *The Documentary History of the Campaign upon the Niagara Frontier in the Year 1812*, part II, (New York, New York: Arno Press & The New York Times, 1971).

Cruikshank, Ernest, *The Documentary History of the Campaign on the Niagara Frontier, 1813*, volume 2, parts I and II, (New York, New York: Arno Press & The New York Times, 1971).

"Letters of Ensign Thomas Warner," The Flag House & Star-Spangled Banner Museum, Baltimore, Maryland.

Lossing, Benson J., *Pictorial Field-Book of the War of 1812*, (New York, New York: Harper and Brothers, 1869).

Heitman, Francis B., *Historical Register and Dictionary of the United States Army From Its Organization, September 29, 1789, to March 2, 1903*, volume I, (Baltimore, Maryland: Genealogical Publishing Company, 1994).

*Historic Markers, Monuments and Memorials of Buffalo, New York, War of 1812*

http://www.andrle.com/markers/mark127.htm

*Public Statutes at Large of the United States of America*, volume II, Congresses 6th through 12th, 1799-1813, (Boston, Massachusetts: Charles C. Little and James Brown, 1845).

Wright, F. Edward, *Maryland Militia, War of 1812, Volume 2, Baltimore*, (Westminster, Maryland: Family Line Publications, 1979).

## "Amongst my best men"

Commodore Isaac Chauncey used the phase "amongst my best men" when describing the 150 Blacks, boys and soldiers he had sent to Master Commandant Oliver Hazard Perry when the ships of the Lake Erie Squadron were being built in 1813. Perry later complained to Chauncey, his commander, that he was sending him untrained sailors. Perry, in fact, had brought Black sailors with him from Rhode Island when he arrived at Erie, Pennsylvania, to take command of the squadron.

It is not known how many Black sailors served with Perry on the Upper Great Lakes during the War of 1812. Estimates range between 10 to 15 percent of the enlisted sailors who fought in the Battle of Lake Erie were Black. This estimate also covers the enlisted personnel serving in the rest of the U.S. Navy during this time period.

The navy did not indicate race on its muster rolls. The names of Black sailors do appear in family and county histories, in books on the War of 1812 and on some pension applications. Only a few names can be verified while most names remain unproven as to their race.

A new source of identifying Blacks who served with the Lake Erie Squadron (plus the rest of the U.S. Navy) lies in the American prisoner of war (POW) records housed in the National Archives of Great Britain. The POW ledgers for the Dartmoor prison not only list the names of the prisoners, how they were captured and when they were sent to this prison, but also their ages, nativity and race. The POW ledgers used in the British colonies, including Canada, did not record age, nativity or race.

A year after the Battle of Lake Erie, four ships from Perry's squadron were captured by the British. The schooners *Ohio* and *Somers* were captured on 12 August 1814 near Fort Erie, across the Niagara River from Buffalo, New York. The schooner *Tigress* was captured on 3 September 1814 while the schooner *Scorpion* was captured on 6 September 1814. Both of these ships were lost to naval actions on Lake Huron.

Eight men had been identified as being Black sailors from the Lake Erie Squadron while assigned to Dartmoor. They are Moses Bailey, Charles Black, Henry Brown, William Griffin, Andrew Norton, Thomas Palmer, John Peters and Jesse Williams. Brown and Williams had served on the U.S. Brig *Lawrence* during the Battle of Lake Erie while Palmer served on the U.S. Schooner *Ariel* and Andrew Norton on the U.S. Brig *Niagara*. Bailey, Black, Griffin and Peters joined the squadron after the battle.

When captured, Griffin, Norton and Palmer were serving on the *Tigress* while Bailey, Brown and Williams were on the *Scorpion*. Black and Peters were on the *Somers*. Bailey died at Dartmoor on 17 Feb 1815 from variola (small pox) while the other men were released from this prison at the end of the war.

All of the men from these ships were first sent to Montreal and then to Quebec before arriving at Halifax, Canada. Black, Griffin and Peters arrived at Quebec from Montreal on 5 October 1814. The other five Black men arrived at Quebec on 1 November 1814. His Majesty's Transport *Freedom* brought Black, Griffin and Peters to Halifax on 1 November 1814 and the other five men on 7 November 1814.

At Halifax, the marines, and the army personnel who were serving as marines, were exchanged or paroled and sent back to the United States. The rest of the captured officers and men were sent to England on H.M. Transport *Argo*.

All of the enlisted men arrived at Dartmoor on 26 Dec 1814, and they were released from the prison on 3 July 1815 (except for Bailey and a few other men who had died at this prison). The officers were sent to a parole station at Dartmouth where they were free to live in the area. They had to abide by strict rules while living in the Dartmouth area. These men did not have to live in the POW prisons.

The seven of the eight Black men from the Lake Erie Squadron were among the last of the American POWs to return home from England after the end of the War of 1812. Many of the Black POWs refused to be released until they were guaranteed that the returning ships would dock at a northern American port. Among the Blacks there were many who were ex-slaves, born free in a northern state, or born free in another country. The fear of becoming a slave or being chained again was too much for many of these men.

| Name | Rank | Age in 1814 | Nativity |
|---|---|---|---|
| Bailey, Moses | Ordinary Seaman | 28 | Pennsylvania |
| Black, Charles | Boy | 18 | Philadelphia |
| Brown, Henry | Cook | 22 | New York |
| Griffin, William | Ordinary Seaman | 23 | New York |
| Norton, Andrew | Ordinary Seaman | 22 | Virginia |
| Palmer, Thomas | Ordinary Seaman | 22 | New York |
| Peters, John | Landsman | 19 | Pennsylvania |
| Williams, Jesse | Ordinary Seaman | 42 | Virginia |

# Bibliography

*General Entry Book of American Prisoners of War* ledger, British Admiralty, Public Record Office, London, Great Britain (ADM 103 / 87 through 91 and 511), Dartmoor ledgers.

*General Entry Book of American Prisoners of War* ledger, British Admiralty, Public Record Office, London, Great Britain (ADM 103 / 362), Quebec ledgers.

*General Entry Book of American Prisoners of War* ledger, British Admiralty, Public Record Office, London, Great Britain (ADM 103 / 640), Dartmoor death certificates.

*General Entry Book of American Prisoners of War* ledger, British Admiralty, Public Record Office, London, Great Britain (ADM 103 / 465, part 2), Quebec transfer rolls and parole location ledgers.

*War of 1812 Service Records - Lake Erie,* Fold3.com, (original data from the National Archives, Washington, DC), excerpts from muster and payroll reports for the Lake Erie Station.

## Some British deaths in Ohio during the War of 1812

During this occasion of the 200th anniversary of the War of 1812, many hereditary societies are in the process of identifying the graves of this war's soldiers and militiamen who are buried in graves throughout Ohio. As part of this effort, new tombstones or medal markers are being installed on many of these graves to identify those men who served during this conflict.

This is a very noble endeavor but we must not forget our enemy, the British soldiers and Canadian militiamen, who fought, died and are buried in Ohio's soil and in the waters of Lake Erie. They are no longer enemies. Many are buried along side Americans. They are now, and forever, Ohioans.

William Jay published a listing of the number of men who were killed during the War of 1812 in a table organized by location and event.[1] He states that 149 Americans and 15 British soldiers were killed at Fort Meigs during the two sieges in 1813. Fort Stephenson had a total of 17 Americans killed and 25 British soldiers killed or died from wounds during the course of the war. The Battle of Lake Erie saw 24 Americans and 70 British soldiers, marines and sailors killed in action and who were buried 'at sea' off Middle Sister Island. However, three American officers and three British officers were buried in a common grave at Put-in-Bay, South Bass Island.

A short distance down the Maumee River from Fort Meigs is the remains of an old British fort, which became the British headquarters during the two sieges. After the first siege the Americans found fresh graves at this old fort, apparently they had been marked since the men knew which were the British graves and which were the American graves.[2]

An outstanding website listing the British causalities during the War of 1812 can be found at www.1812casualties.org. This online database lists the members of the Royal Army, Royal Navy, Royal Marines, the Canadian militia, and the Provincial Marine who were killed in action, died of wounds, died from other causes, who deserted or who became prisoners of war in the United States.

With the permission of the website owners, the listing of 71 British soldiers and Canadian militiamen who died in Ohio can be found after this article. Still to be identified and listed, are the British army and naval personnel who were prisoners of war and who died at the prisoner of war camp outside of Chillicothe, Ohio; also the British naval deaths from the Battle of Lake Erie.

The men served in the 41st Regiment of Foot, the Royal Newfoundland Regiment of Fencible Infantry, and in two Essex County militia regiments. The 41st Regiment of Foot was organized as the Colonel Edmund Fielding's Regiment of Invalids on 1 March 1719.[3] At the outbreak of the War of 1812, the 41st Regiment of Foot was the only full British regiment operating in Upper Canada (now Ontario). This regiment would become the most decorated British regiment during the War of 1812.

The Royal Newfoundland Regiment was raised in Newfoundland on 25 April 1795 to defend the colonies in British North America during the Napoleonic Wars. Fencibles were regular soldiers raised for local duty and who could not be used for overseas service.[4] This regiment served in Upper Canada during most of the war.

---

[1] *Collections of the New York Historical Society*, second series, volume II, (New York: 1849), Table of the Killed and Wounded in the War of 1812, by William Jay.

[2] Brown, Samuel R., *Views of the Campaigns of the North-Western Army*, (Philadelphia, Pennsylvania: Griggs and Dickinson, Printers, 1815), page 115.

[3] Lomax, David Alexander Napier, *History of the Services of the 41st (the Welsh) Regiment*, (Devonport : Printed by Hiorns & Miller, army printers, 1899).

[4] Hitsman, J. Mackay, updated by Donald E. Graves, *The Incredible War of 1812, A Military History*, (Robin Brass Studio, Toronto, Canada: 1999), page 322.

Militiamen from the 1st and 2nd Essex County Militia Regiments were called for military duty to support the British Army during this war.[5] This county is the southernmost county in Canada and it borders Detroit, Michigan, across the Detroit River.

There are two memorials to the fallen British in Ohio. The International Peace Memorial on South Bass Island has a crypt at its base where the remains of the three American and three British officers who died during the Battle of Lake Erie were laid to rest. The British officers were Robert Finnis, John Garland and James Garden while the American officers were John Brooks, Henry Laub and John Clark.

In Fremont, Ohio, there is a plaque commemorating the British soldiers who died during the Battle of Fort Stephenson. The plaque is mounted on the outside wall of the public school building near where the men were buried. The following words are on the plaque:

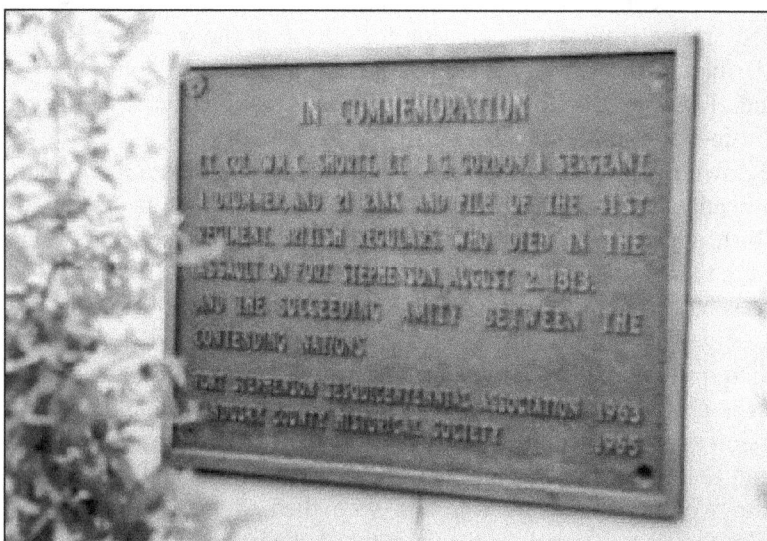

In Commemoration - Lt. Colonel Wm. C. Shortt, Lt. J. G. Gordon, 1 sergeant, 1 drummer, and 21 rank and file of the 41st Regiment, British regulars, who died in the assault on Fort Stephenson, August 2, 1813, and the succeeding amity between the contending nations. Erected by Fort Stephenson Sesquicentennial Association 1963 and Sandusky County Historical Society 1965.

KIA – Killed in action
DOW – Died from wounds
Lt – Lieutenant
Capt - Captain

**Essex County Militia, Ontario**

| Name | Rank | Company Officer | Manner of Death | Place of Death | Date of Death |
|------|------|-----------------|-----------------|----------------|---------------|
| Bondy, Laurente | Captain | Not Applicable | KIA | Fort Meigs | 5 May 1813 |
| Roberts, William | Private | Unknown | DOW | Fort Meigs | 10 May 1813 |
| Dufour, Jean Baptiste | Private | Capt Alexis Maisonville | Died | Fort Meigs | 1 May 1813 |

---

[5] Irving, L. Homfray, *Officers of the British Forces in Canada during the War of 1812*, (Welland Tribune Printers, 1908).

**41st Regiment of Foot - 1st Battalion**

| Name | Rank | Company Officer | Manner of Death | Place of Death | Date of Death |
|------|------|-----------------|-----------------|----------------|---------------|
| Ancliffe, Thomas | Private | Lt Thomas Martin | KIA | Lake Erie | 10 Sep 1813 |
| Baggan, James | Private | Capt William L. Crowther | KIA | Lake Erie | 10 Sep 1813 |
| Barkley, James | Private | Capt Joseph Tallon | KIA | Fort Meigs | 5 May 1813 |
| Bedman, William | Private | Capt Joseph Tallon | KIA | Lake Erie | 10 Sep 1813 |
| Bellows, George | Private | Capt Angus McIntyre | KIA | Fort Stephenson | 2 Aug 1813 |
| Bird, Edward | Private | Capt William L. Crowther | KIA | Fort Stephenson | 2 Aug 1813 |
| Bishop, John | Drummer | Capt Joseph Tallon | KIA | Fort Stephenson | 2 Aug 1813 |
| Booth, Richard | Private | Capt Richard Bullock | KIA | Fort Meigs | 5 May 1813 |
| Bow, James | Private | Lt Thomas Martin | KIA | Lake Erie | 10 Sep 1813 |
| Burge, Thomas | Private | Capt Angus McIntyre | KIA | Fort Stephenson | 2 Aug 1813 |
| Butson, William | Private | Capt William L. Crowther | KIA | Fort Stephenson | 2 Aug 1813 |
| Carpenter, Nathaniel | Private | Capt Joseph Tallon | KIA | Lake Erie | 10 Sep 1813 |
| Carpmail, William | Private | Capt Joseph Tallon | KIA | Fort Meigs | 5 May 1813 |
| Cartledge, Samuel | Private | Capt Adam Muir | KIA | Fort Meigs | 5 May 1813 |
| Chamberlane, John | Private | Capt Adam Muir | DOW | Fort Meigs | 20 May 1813 |
| Cokely, Cornelius | Private | Lt John H. JeBoult | KIA | Lake Erie | 10 Sep 1813 |
| Collison, William | Private | Lt Thomas Martin | KIA | Fort Stephenson | 2 Aug 1813 |
| Cork, William | Private | Capt William L. Crowther | KIA | Fort Stephenson | 2 Aug 1813 |
| Cox, John | Private | Capt Richard Bullock | KIA | Fort Meigs | 5 May 1813 |
| Currant, James | Private | Capt William L. Crowther | DOW | Fort Stephenson | 25 Aug 1813 |
| Dawdle, Richard | Private | Capt William L. Crowther | DOW | Fort Stephenson | 3 Aug 1813 |
| Dorman, Benjamin | Private | Capt Joseph Tallon | KIA | Fort Meigs | 5 May 1813 |
| Downs, Thomas | Private | Capt Joseph Tallon | DOW | Fort Meigs | 1 Jun 1813 |
| Dunford, Isaac | Private | Capt Angus McIntyre | KIA | Fort Stephenson | 2 Aug 1813 |
| Dyer, John | Private | Capt Adam Muir | KIA | Fort Meigs | 5 May 1813 |
| Elsby, James | Private | Lt Thomas Martin | KIA | Fort Stephenson | 2 Aug 1813 |
| Fisher, Joseph | Corporal | Capt Angus McIntyre | KIA | Fort Stephenson | 2 Aug 1813 |
| Frith, Samuel | Private | Lt John H. JeBoult | KIA | Lake Erie | 10 Sep 1813 |
| Golding, George | Private | Lt John H. JeBoult | KIA | Lake Erie | 10 Sep 1813 |
| Gordon, James George | Lieutenant | Not Applicable | KIA | Fort Stephenson | 2 Aug 1813 |
| Graves, Edward | Private | Capt Joseph Tallon | KIA | Fort Meigs | 5 May 1813 |
| Lander, James | Sergeant | Capt Angus McIntyre | KIA | Fort Stephenson | 2 Aug 1813 |
| Lee, Robert | Private | Lt William L. Crowther | KIA | Fort Stephenson | 2 Aug 1813 |
| Leonard, John | Private | Lt John H. JeBoult | KIA | Lake Erie | 10 Sep 1813 |
| McDevitt, Daniel | Corporal | Lt Richard Bullock | KIA | Fort Stephenson | 2 Aug 1813 |
| Middleton, Abner | Private | Capt Angus McIntyre | DOW | Fort Stephenson | 3 Aug 1813 |
| Pamken, James | Private | Lt Thomas Martin | DOW | Fort Stephenson | 30 Aug 1813 |
| Pomroy, Thomas | Private | Capt Angus McIntyre | DOW | Fort Stephenson | 4 Aug 1813 |
| Prescott, John | Private | Capt William L. Crowther | KIA | Fort Stephenson | 2 Aug 1813 |
| Ratcliffe, John | Private | Lt Richard Bullock | KIA | Lake Erie | 10 Sep 1813 |
| Reeves, William | Corporal | Capt William L. Crowther | KIA | Fort Stephenson | 2 Aug 1813 |
| Russell, Patrick | Private | Lt John H. JeBoult | KIA | Fort Meigs | 5 May 1813 |
| Sasse, Henry | Private | Lt Richard Bullock | KIA | Fort Stephenson | 2 Aug 1813 |

## 41st Regiment of Foot - 1st Battalion (Continued)

| Name | Rank | Company Officer | Manner of Death | Place of Death | Date of Death |
|---|---|---|---|---|---|
| Shakleton, Robert | Private | Capt Joseph Tallon | DOW | Fort Meigs | 1 Jun 1813 |
| Shanahan, John | Private | Lt Richard Bullock | KIA | Fort Stephenson | 2 Aug 1813 |
| Shortt, William Charles | Brevet Lieutenant Colonel | Not Applicable | KIA | Fort Stephenson | 2 Aug 1813 |
| Smith, Giles | Private | Lt Richard Bullock | KIA | Fort Stephenson | 2 Aug 1813 |
| Wardle, John | Private | Lt Richard Bullock | KIA | Fort Stephenson | 2 Aug 1813 |
| Watkins, Thomas | Private | Lt John H. JeBoult | KIA | Lake Erie | 10 Sep 1813 |
| Westlake, James | Private | Capt William L. Crowther | KIA | Fort Stephenson | 2 Aug 1813 |
| White, James | Private | Capt Angus McIntyre | KIA | Fort Stephenson | 2 Aug 1813 |
| Williams, Thomas | Private | Capt Joseph Tallon | KIA | Fort Stephenson | 2 Aug 1813 |

## Royal Newfoundland Regiment of Fencible Infantry

| Name | Rank | Company Officer | Manner of Death | Place of Death | Date of Death |
|---|---|---|---|---|---|
| Brennan, Michael | Private | Capt Robert Mockler | KIA | Lake Erie | 10 Sep 1813 |
| Brien, William | Private | Capt John T. Whelan | DOW | Lake Erie | 10 Sep 1813 |
| Deaton, Thomas | Private | Capt Tito F. Lelievre | KIA | Fort Meigs | 28 May 1813 |
| Deolin, Patrick | Private | Unknown | KIA | Fort Meigs | 5 May 1813 |
| Fahey, Thomas | Private | Capt John T. Whelan | KIA | Lake Erie | 10 Sep 1813 |
| Garden, James | Lieutenant | Not Applicable | KIA | Lake Erie | 10 Sep 1813 |
| Gardiner, William | Private | Capt Robert Mockler | KIA | Lake Erie | 10 Sep 1813 |
| Hawkins, Richard | Private | Capt John T. Whelan | KIA | Lake Erie | 10 Sep 1813 |
| Louis, Donald | Private | Capt Tito F. Lelievre | KIA | Lake Erie | 10 Sep 1813 |
| Mansfield, Robert | Private | Capt John T. Whelan | KIA | Lake Erie | 10 Sep 1813 |
| O'Neil, Gregory | Private | Capt John T. Whelan | KIA | Lake Erie | 10 Sep 1813 |
| Riley, Patrick | Private | Unknown | DOW | Lake Erie | 10 Sep 1813 |
| Smith, John | Private | Capt Robert Mockler | KIA | Lake Erie | 10 Sep 1813 |
| Tinham, Thomas | Private | Capt Robert Mockler | KIA | Lake Erie | 10 Sep 1813 |
| Wilcox, Daniel | Sergeant | Capt Robert Mockler | Died | Lake Erie | 10 Sep 1813 |
| Woodlands, William | Sergeant | Capt Tito F. Lelievre | KIA | Lake Erie | 10 Sep 1813 |

# United States Corps of Artificers

The United States Corps of Artificers was created by Congress on 23 April 1812 in order to built and maintain the wagons and boats of the Quartermaster General's Department of the U.S. Army.[6] The company-side corps was under the command of Superintendent Alexander Parris and it was attached to the Quartermaster General's Department.

The corps had an authorized strength of 130 men, made up of both army personnel and civilian workers. They had their own distinct uniform. The superintendent had the equivalent rank of a captain in the army. The corps was created for a period of three years, unless disbanded sooner by the President of the United States. The corps only operated with the Army of the North in New York and Vermont.

### Corps of Artificers

| | |
|---|---|
| 1 superintendent | 20 house carpenters |
| 4 assistants | 5 ship carpenters |
| 2 master masons | 20 blacksmiths |
| 2 master carpenters | 16 boat-builders |
| 2 master blacksmiths | 16 armorers |
| 2 master boat-builders | 12 saddle & harness makers |
| 2 master armorers | 24 laborers |
| 2 master saddle and harness makers | |

The corps was a construction organization which could build and repair supply depots, wagons and boats of the Quartermaster's General Department. The unit could also make and repair saddles and harness for the department's horses. The blacksmiths could make and repair iron tools and equipment plus horse shoes.

A muster roll of the corps for May-June 1814, while the corps was stationed at the Burlington Cantonment in Burlington, Vermont, shows the manpower strength of only 73 men.[7] Superintendent Parris, while he was in Boston, Massachusetts, in November 1814, wrote a letter to the commander of the Army of the North, Major General Henry Dearborn, asking the general if he could continue to recruit men for the corps.[8] At this time, the corps only had 68 men assigned. It appears that the corps was never at full strength.

The muster roll for May-June 1814 is a snap-shot in time. The army conducted muster rolls six times a year on the last day of even numbered months. The recruiting reports and muster rolls for the years 1812 through 1815 would need to be found and combined in order to tabulate the total number of men who were assigned to this corps.

Researching each man who was found in the May-June 1814 muster, by using the army's *Register of Enlistments*,[9] pension applications [10] and the military land bounty warrants [11] & patents [12], revealed more

---

[6] *Public Statutes at Large of the United States of America*, volume II, (Boston: Charles C. Little and James Brown, 1845), Twelfth Congress, Session 1, Chapter LIX, page 710, 23 April 1812, "an Act for the organization of a Corps of Artificers."

[7] Clark, Byron N., *A List of Pensioners of the War of 1812*, (Research Publication Company: Burlington and Boston, 1904), Payroll of a company of artificers, commanded by Alexander Parris, pp. 129-131.

[8] Fredriksen, John C., *The War of 1812: U.S. War Department Correspondence, 1812-1815*, (McFarland & Company, Inc.: Jefferson, NC 2016), page 296.

[9] *Register of Enlistments in the U.S. Army, 1798-1914*; (Washington, DC: National Archives Microfilm Publication M233); Records of the Adjutant General's Office, 1780's-1917, Record Group 94.

[10] *War of 1812 Pension Applications*; National Archives Microfilm Publication M313; Records of the Department of Veterans Affairs, Record Group Number 15; National Archives, Washington D.C.

details about this corps. Twenty-four men were found as having been detached from their army regiments in order to serve in this corps. In the service records of these men it is stated that they were transferred to the Corps of Artificers but they received their land bounties and pensions through their army regiments.

Additional twelve men were found to have served in this corps while the author was researching the muster roll. Thirteen men were listed as "Black," all serving as laborers. Some of these Blacks were army soldiers who were detached from their regiments.

The rest of the 49 men were probably civilians who did not qualify for land bounties and pensions. Most of the army personnel assigned to the corps did not serve the entire three years and ten days that the corps was in existence. Many of the army soldiers had enlistment periods of one-year or eighteen months.

From information on the affidavits in the pension applications of some of the men, it can be determined where the corps served during the War of 1812. It appears that Parris started recruiting men from the Boston, Massachusetts, area before the corps joined the Army of the North. Some of the affidavits stated that the corps was attached to the 4th Regiment of U.S. Infantry.

The corps wintered over at the Greenbush Cantonment outside of Albany, New York, during the winter of 1812-1813. They were stationed at Fort George, Upper Canada, after the Americans captured this fort in May 1813.

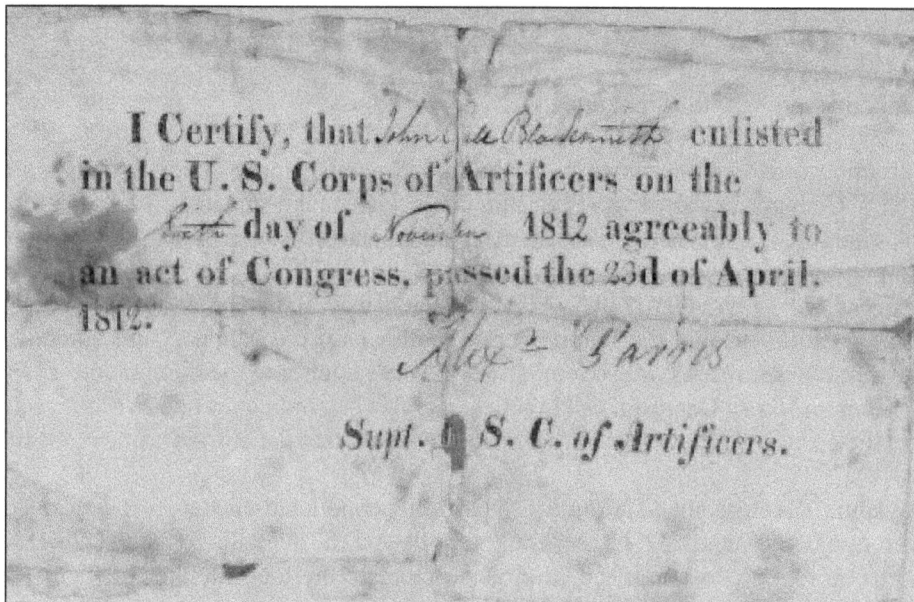

A certificate of enlistment for the U.S. Corps of Artificers

At some point in time, the corps was stationed at Sackets Harbor, New York. They were serving at Plattsburgh, New York, in February 1814 and then stationed at the Burlington Cantonment in Burlington, Vermont, in May-June 1814. They were once again at Plattsburgh during and after the Battle of Plattsburgh in September 1814.

Congress passed the act to establish the peacetime army on 3 Mar 1815.[13] The Corps of Artificers was not included among the regiments and corps that were being retained in the service and the corps was

---

[11] *War of 1812 Military Bounty Land Warrants, 1815-1858*; National Archives Microfilm Publication M848; Records of the Bureau of Land Management, Record Group 49; National Archives, Washington, D.C.

[12] United States. Bureau of Land Management, General Land Office Records. Automated Records Project; *Federal Land Patents*, State Volumes. http://www.glorecords.blm.gov/. Springfield, Virginia: Bureau of Land Management, Eastern States, 2007.

[13] *Public Statutes at Large of the United States of America*, volume III, (Boston: Charles C. Little and James Brown, 1846), Thirteenth Congress, Session III, Chapter LXXIX, pp. 224-225, 3 March 1815, "an act fixing the military peace establishment of the United States."

disbanded. Most of the soldiers in the corps were discharged at Greenbush on this date.

**Payroll of a company of artificers, commanded by Alexander Parris, superintendent of the Corps of Artificers of the United States for the months of May and June, 1814.**

Andress, Samuel N. - Carpenter

Baptist, John - Laborer - Listed as "Black" in *A List of Pensioners of the War of 1812.*

Bemont, Edmund - Laborer - Discharged on 1 Jun 1814.

Billings, Zebina - Carpenter - Enlisted: 21 May 1813 - Period: 1 Year - Enlisted in the 29th Infantry and transferred to the Corps of Artificers, discharged on 14 Jun 1814.

Bronsdon, Joseph R. - Master Carpenter

Buck, Elias - Blacksmith - Discharged on 1 Jun 1814.

Bush, Stephen - Carpenter - Age: 25 - Height: 5' 8" - Born: Saddle River, NJ - Trade: Carpenter - Enlisted: 26 May 1813 - Period: War - enlisted in the 29th Infantry, transferred to Corps of Artificers on 25 Mar 1814, discharged on 26 May 1814.

Callender, Henry B. - Saddler

Campfield, Jesse - Carpenter

Chase, Nathaniel - Master Carpenter - Discharged on 1 Jun 1814.

Clinton, Thomas - Carpenter - Discharged on 1 Jun 1814.

Cobb, Cyrus - Master Mechanic

Crapin, Samuel - Laborer - Enlisted: 8 Feb 1813 - Period: War - BLW 26800-160-12 - attached to the Corps of Artificers, waiter to Ira Floyd, superintendent of Corps of Artificers, discharged on 8 May 1815.

Damon, Zachariah - Carpenter

Derby, Abraham - Laborer - Enlisted: 24 Mar 1813 - Period: 18 Months - enlisted in the 21st Infantry, transferred to Corps of Artificers on 26 May 1814, listed as "Black" in *A List of Pensioners of the War of 1812.*

Dibble, Asa - Laborer - Enlisted: 7 Jan 1813 - Period: 5 Years - Pension: Land bounty to John Dibble and other heirs at law of Asa Dibble, deceased - BLW 20244-160-12 - enlisted in 23rd Infantry, attached to Corps of Artificers on 18 Jul 1813; died on 15 Jan 1815.

Dunn, Joseph - Master Artisan – Enlisted : 1 Oct 1812 - Discharged on 2 May 1815 at Plattsburgh, NY – Bounty : BLW 56476-80-50, BLW 23094-80-55 – Pension : Wife Susan W. Moses, WO-6332, WC-3131, Captain Alexander Paris' Company, Maine Militia, married on 9 Sep 1810 in Portland, ME, soldier died on 1 Apr 1846 in New York, NY.

Dunn, Joshua - Saddler - Enlisted: 11 Oct 1812 - Pension: Wife Alice H. (second wife), SO-915, SC-386, WO-40333, WC-30899; married on 27 Jun 1879, Portland, Cumberland County, ME; soldier died on 20 Aug 1880 in Portland, ME; Captain Alexander Parris' Company, US Artificers - BLW 454-160-50 - discharged on 2 May 1815 at Plattsburg, NY.

Eaton, Jacob F. - Armorer - Age: 25 - Height: 5' 10" - Trade: Blacksmith - Enlisted: 2 Feb 1813 - Period: 5 Years - BLW 26927-160-12 – Enlisted in the 11th Infantry, attached to Corps of Artificers on 4 Jul 1814, discharged on 2 May 1815.

Eaton, William - Laborer

Emery, Joseph - Boat Builder - Age: 34 - Height: 5' 10" - Trade: Cooper - Enlisted: 15 Feb 1813 - Period: War - BLW 20376-160-12 - Enlisted in the 11th Infantry, attached to Corps of Artificers on 3 Mar 1814, discharged on Plattsburgh, NY, on 3 May 1815.

Floyd, Ira - Assistant Superintendent

Follenbee, Timothy - Ship Carpenter

Fuller, Issachar - Saddler - Enlisted: 27 Oct 1812 - Pension: Wife Matilda C. Nichols, WO-21049, WC-11915; married on 28 Jan 1822, Cohasset, Norfolk County, MA; soldier died on 11 Feb 1866 in Hingham, MA; soldier in Captain Alexander Parris' Company, US Army Corps of Artificers - BLW 77579-40-50; BLW 34927-120-55 - discharged on 31 Jan 1813.

Gragg, Enus - Ship Carpenter

Hartshorn, David - Carpenter

Hatch, Ward - Laborer

Hatford, Samuel - Laborer - Listed as "Black" in *A List of Pensioners of the War of 1812*.

Hearn, Daniel - Laborer

Herndon, John - Carpenter

Hotchkiss, Henry - Master Blacksmith - Age: 26 - Height: 5' 10" - Born: Berlin County, VT - Trade: Blacksmith - Enlisted: 21 Jul 1812 - Period: 5 Years - Pension: Wife Sarah Dorrance Cochran, WO-3895, WC-4744; married on 30 Oct 1806, Durham, NY; soldier died on 21 Nov 1871 in Durham, NY - BLW 27061-160-12, BLW 162101-50 Cancelled (entitled to 160 acres of land) - Enlisted in the 23rd Infantry, promoted to Sergeant Major on 21 Jul 1812, attached to Corps of Artificers in Oct or Nov 1812, discharged on 15 Jan 1813, injury to his eyes.

Houtzell, Jacob - Armorer

Howland, Elisha W. - Carpenter

Josselyn, Isaac - Laborer - Discharged on 1 Jun 1814.

Lane, George - Carpenter

Lawrence, Nathaniel - Laborer - Listed as "Black" in *A List of Pensioners of the War of 1812*.

Marshall, Samuel W. - Laborer - Listed as "Black" in *A List of Pensioners of the War of 1812*.

Martin, John - Carpenter

Melona, Cornelius - Laborer - Listed as "Black" in *A List of Pensioners of the War of 1812*.

Millikin, James - Laborer - Listed as "Black" in *A List of Pensioners of the War of 1812*.

Millikin, Sterling F. - Saddler - Age: 24 - Height: 5' 8" - Born: North Hampton, NC - Trade: Saddler - Enlisted: 24 Mar 1812 - Period: 5 Years - BLW 20747-160-12 - Enlisted in the 5th Infantry, attached to the Corps of Artificers after August 1814, discharged on 3 May 1815.

Nichols, Josiah - Armorer - Age: 42 - Height: 5' 10" - Born: Mansfield, CT - Trade: Armorer - Enlisted: 20 Feb 1813 - Period: War - BLW 3937-160-12 - Enlisted in 11th Infantry, transferred to Corps of Artificers on 4 Jul 1814, discharged on Burlington, VT, on 1 Jun 1815.

Noyes, Jacob - Carpenter - Pension: Wife Maria, SO-24668, SC-21804, WO-43102, WC-33883, Captain Alexander Parris' Company, US Artificers.

Page, Charles - Laborer - Discharged on 1 Jun 1814.

Parker, James - Carpenter

Parris, Alexander - Superintendent – commissioned on 17 Jun 1812 and discharged in Jun 1815.

Payson, Edward - Saddler - Discharged on 1 Jun 1814.

Phillips, Christopher - Ship Carpenter

Pierce, John - Carpenter - Pension: Wife Abigail W., SO-5602, SC-3141, WO-16473, WC-14517, Captain Alex Paris's Company, US Artificers - Discharged on 11 Jul 1814.

Pinkham, Charles - Carpenter - Enlisted: 30 May 1813 - Period: 1 Year - BLW 1135-160-12 - Enlisted in the 34th Infantry, transferred to Corps of Artificer, discharged on 30 May 1815.

Pudney, Justis - Blacksmith - Discharged on 1 Jun 1814.

Richardson, Edward - Carpenter

Ridgeway, Samuel - Laborer - Listed as "Black" in *A List of Pensioners of the War of 1812*.

Pool, Galen – Carpenter – BLW 26687-160-50, Captain Paris' Company, U.S. Artificers

Sawyer, Nathaniel - Laborer - Listed as "Black" in *A List of Pensioners of the War of 1812*.

Siniler, John C. - Saddler

Smith, William N. - Carpenter

Sprague, Frederick A. - Carpenter - Pension: Wife Bridget, WO-32501, Captain Alexander Parris, US Artificers - BLW 12725-160-50.

Storrs, Stephen - Laborer - Enlisted: 13 Jan 1813 - Period: 5 Years - BLW 24581-160-12 – Storrs affidavit states that he was captured by the British during the Battle of Crystler's Field, Upper Canada, on 26 Oct 1813 and was exchanged at Chazy, New York, on 11 May 1814; he was made a prisoner of war again during the Battle of Lewisburg in New York, he was released and then attached to the Corps of Artificers.

Sturtivant, Martin - Armorer - Discharged on 1 Jun 1814.

Sullivan, Timothy - Carpenter - Enlisted: 12 Jun 1813 - Period: 1 Year – Enlisted in the 29th Infantry, attached to Corps of Artificers, discharged on 12 Jun 1814.

Taggart, James - Boat Builder - Age: 47 - Height: 5' 8" - Born: Goffstown, NH - Trade: Carpenter - Enlisted: 26 Apr 1813 - Period: War - BLW 11669-160-12 – Enlisted in the 4th Infantry, attached to the Corps of Artificers on 3 Mar 1814, discharged at Greenbush, NY, on 18 May 1815.

Thomas, Edward - Carpenter

Thomas, Moses - Carpenter - Discharged on 1 Jun 1814.

Thomas, Robert - Laborer - Listed as "Black" in *A List of Pensioners of the War of 1812*.

Townsend, Jonathan - Laborer - Age: 25 - Height: 5' 6" - Born: N. Salem, Hampshire County, MA - Trade: Blacksmith - Enlisted: 19 Feb 1813 - Period: 5 Years – Enlisted in the 9th Infantry, attached to the Corps of Artificers at Plattsburgh on 1 May 1814, transferred to 5th Infantry, deserted at Detroit, MI, on 26 Jun 1816, listed as "Black" in *A List of Pensioners of the War of 1812*.

Waldron, Richard - Carpenter - Born: Prince George County, MD - Enlisted: 20 Apr 1813 - Place: Baltimore, MD - Period: 5 Years - Pension: Old War IF-26802, artificer, Captain Parish, USA – Enlisted in the 5th Infantry, attached to Corps of Artificers at Plattsburg, NY, on 1 Aug 1814; transferred to Corps of Artificers; discharged on Surgeon's Certificate in Jul 1814 due to injury in the line of duty at Plattsburg.

Waterhouse, Jacob - Master Blacksmith

Wheelock, Ezra - Laborer - Age: 25 - Height: 5' 4" - Born: Worcester, MA - Trade: Blacksmith - Enlisted: 9 Feb 1813 - Place: Worcester, MA - Period: War - BLW 11725-160-12 – Enlisted in the 4th Infantry, on extra duty in Quartermaster General's Department, 28 Feb 1815; transferred to the 5th Infantry, discharged at Greenbush, NY, on 1 May 1815; listed as "Black" in *A List of Pensioners of the War of 1812.*

Whicher, Samuel E. - Laborer

Whicher, William - Laborer

Whitney, Josiah - Saddler – Enlisted : 1 Mar 1814 - Pension: Wife Almira, WO-13146, WC-7193, married on 28 May 1818 at Rootstown, Portage County, OH, soldier died on 7 Oct 1869 at Rootstown, Captain Alexander Parris, US Corps of Artificers - BLW 31860-80-50, BLW 12343-80-55 - Discharged on 1 Jun 1814 – also served from 19 Feb 1813 to 31 Jul 1813

Wilkinson, John - Carpenter

Williams, Ebenezer - Carpenter - Discharged on 5 Jun 1814.

Wiman, Ebenezer (Wyman) - Laborer - Listed as "Black" in *A List of Pensioners of the War of 1812.*

Wright, Obediah - Assistant Superintendent

## Additional members of the Corps of Artificers

Barker, Ira – Artificer – BLW 20312-160-12, Captain Parris' Corps, U.S. Artificers – submitted an affidavit stating that he served with Lewis Litchfield in the Corps of Artificers.

Bates, Ward – Artificer – BLW 51199-80-50, Captain Parris' Company of Artificers, BLW 15520-80-50, Captain Parris' Company, Massachusetts Militia.

Estes, Richard – Private – BLW 12043-160-12, Captain Henry Phillips' Company, 6th Infantry, – submitted an affidavit stating that he served with Lewis Litchfield in the Corps of Artificers.

Davis, Ebenezer – Private – BLW 45731-80-55, land bounty to Betsey McLellan, former widow of Ebenezer Davis, Captain Alexander Parris' Company, US Artificers.

Fish, Zachariah - Carpenter -BLW 89304-160-55, land bounty to Luckey Fish, widow of Zachariah Fish, Captain Parris' Corps of Artificers.

Gale, Ephraim B. – Artificer – Enlisted : 1 Dec 1812 – Discharged on 4 Feb 1814 – Bounty : BLW 19758-160-50 – Pension : Wife Betsey Proctor, SO-8380, SC-8609, Captain Parris' Company, Massachusetts Militia – married in May 1814 at Haverhill, MA – soldier died about 1875, brother to John Gale.

Gale, John - Blacksmith – Enlisted : 1 Nov 1812 - Pension: Wife Mary R. Colby, SO-12409, SC-16763; married on 5 Jul 1818, Haverhill, MA; soldier died on 19 Feb 1877, artisan, Captain Alexander Paris' Company, US Engineers - BLW 24150-160-50 - discharged on 31 Jul 1813, brother to Ephraim B. Gale.

Benson, Consider – Private – BLW 16296-160-55, Captain Parris' Company, Massachusetts Militia.

Hollowell, Benjamin G. – Private – BLW 55676-160-55, Captain Parris' Company, Massachusetts Militia.

Litchfield, Lewis - Private - Pension: Wife Betsey Stetson, SO-28533, SC-20408, WO-31221, WC-16075, Captain A. Parris' Company, US Artificers. BLW 26168-160-50, BLW 50929-80-50 Cancelled (entitled to 160 acres of land )– Enlisted : 1 Jan 1813 – Discharged 31 Mar 1813 for disability, married on 14 Dec 1851 (both previous married), soldier died on 5 Jun 1871 in Hanover, MA.

Nichols, William – Pension: Wife Isabella L., WO-30058, WC-18243, Captain A. Parris' Company, Corps of Artificers, USA.

Partridge, Flavel – Pension: SO-17874, SC-11185, Captain Alexander Parris' Company, US Artificers.


## Penn State Erie honors the memory of the Battle of Lake Erie

Walking the campus of Pennsylvania State University at Erie, Pennsylvania, can give you an eerie feeling especially at night. All of the residence dormitories are named after the prominent men and their ships that participated in the Battle of Lake Erie on 10 September 1813.

Of the 4,000 plus students enrolled in this university, nearly 1,650 live in these dormitories, which honor the memory of a famous naval battle which was fought during the War of 1812.

Lawrence Hall is named after the United States Brig *Lawrence*, which was launched on 25 June 1813 as the flagship of Master Commandant Oliver Hazard Perry's Lake Erie Squadron. Lieutenant John Yarnall commanded this Ship.

The sister ship of the *Lawrence* was the U.S. Brig *Niagara* which graces the name of the next dormitory at Penn State Erie. This brig was launched on 4 June 1813 and it was commanded by Master Commandant Jesse D. Elliott. A replicate of this ship sails the Great Lakes representing Erie and the Commonwealth of Pennsylvania. Both the *Lawrence* and the *Niagara* were built in Erie.

Perry Hall is named after Master Commandant Perry. After the battle he was promoted to captain. Ohio Hall is named after the one-gun schooner which was built in Cleveland, Ohio. This ship was commanded by Sailing Master Daniel Dobbins.

Porcupine Hall was named after the one-gun schooner of the same name. It was commanded by Sailing Master George Senat. Tiffany Hall was named for Cyrus Tiffany, an African-American who served on board the Niagara during the battle.

Tigress Hall was name after the one-gun schooner Tigress. This schooner was originally named the *Amelia*. The *Tigress* was commanded by Lieutenant Augustus H.M. Conklin.

Turner Hall was named after Lieutenant Daniel Turner, who commanded the U.S. Brig *Caledonia*. Caledonia Hall was named after the third brig in the squadron.

Packett Hall was named after Lieutenant John Packett who commanded the clipper-built schooner *Ariel*. Ariel Hall was named after the four-gun schooner that was commanded by Packett.

Champlin Hall was named after Sailing Master Stephen Champlin, who commanded the schooner *Scorpion*. Scorpion Hall was named after this schooner.

Elliot Hall was named after Master Commandant Jesse D. Elliot, the ranking officer below Perry. He commanded the *Niagara* during the battle.

Yarnall Hall was named for Lieutenant John J. Yarnall, the first lieutenant aboard the *Lawrence*. Somers Hall was named after the two-gun schooner of the same name.

Senat Hall is named for Acting Sailing Master George Senat, who commanded the one gun schooner *Porcupine*. Almy Hall was named after Thomas C. Almy, who was the Sailing Master of the schooner *Somers*.

# The Privateers

The most successful arm of the naval forces of the United States during the War of 1812 was the privateers. The ships and the men were a part of the American merchant fleet, which obtained Letters of Marque and Reprisal from the federal government in order to prey on British and Canadian shipping. The privateers were very similar to the militia forces on land. They were regulated and controlled by the U.S. Congress.

The Act of 26 June 1812 authorized the President to grant and revoke all Letters of Marque and Reprisal through the Secretary of State.[14] In applying for a Letter, the owner had to fill out an application stating the name and description of the vessel along with the tonnage and arms, the owner of the vessel, number of crew. A bond of $5,000 was posted for vessels with a crew of less than 150 and $10,000 for vessels with a crew of over 150 men.

All prizes, captured enemy vessels and cargo, were turned over to the district courts of the United States where they were auctioned off and the money divided between the owners of the vessels and the crew. Recaptured American vessels were to be handled back to their original owners.

The privateers were required to maintain a journal of their cruising activities, which was turned over to the customs officer upon completion of each cruise. The journals were then given to the district courts. All prisoners captured by the privateers were turned over to the district marshal or a military officer and a bounty of $20 per prisoner was added to the prize money. By the end of the war, this bounty had been increased to $100 per prisoner.

There were stiff penalties for any owner or commander of a privateer who violated any act of Congress. Their Letter of Marque and Reprisal could be forfeited. Any officer or seamen on privateers who committed a crime were subject to a court-martial, which was made up of officers from at least two different privateers or by naval officers of the U.S. Navy. The Secretary of the Navy was notified of all court-martials.

Two percent of the prize money, minus all charges and expenditures, was turned over to the federal government and placed in the fund to support the wounded and the widows and orphans of deceased officers and seamen of privateers.

Congress continued to regulate the privateers throughout the War of 1812. The Act of 27 January 1813 required that all prizes and captured property were to be sold at public auctions within 60 days by the district marshal.[15] Duties were to be paid on all goods captured from the enemy.

The Act of 13 February 1813 required that the money collected for the privateer's pension fund was to be turned over to the U.S. Treasury and the officers and crew of a privateer who was wounded or disabled due to any engagement were to be placed on the navy's pension list.[16] The captains of privateers were now required to maintain a record in their journals of those who were wounded or injured.

The heirs of any officer or seamen who had died in the line of duty on a privateer since the start of the war were entitled to receive half pay for five years under the Act of 4 March 1814.[17] The names of those men were added to the navy's pension list.

A total of 517 American merchant ships were armed and received a Letter of Marque and Reprisal during the War of 1812. The armed vessels captured 1,300 enemy vessels, 6,000 prisoners, and goods

---

[14] *Twelfth Congress, Session I, Chapter 107, 26 June 1812, pages 759-764*, An act concerning Letters of Marque, Prizes and Prize Goods

[15] *Twelfth Congress, Session II, Chapter 13, 27 January 1813, pages 792-793*, An act in addition to the act concerning Letters of Marque, Prizes and Prize Goods

[16] *Twelfth Congress, Session II, Chapter 22, 13 February 1813, pages 799-800*, An act regulating pensions to persons on board private armed ships

[17] *Thirteenth Congress, Session II, Chapter 20, 4 March 1814, pages 103-104*, An act giving pensions to the orphans and widows of persons slain in the public or private armed vessels of the United States

valued at $40,000,000. These ships were not the merchant marine ships that were used during the 20th century to transport troops and supplies during a war. These ships were privately owned warships, which supplemented the U.S. Navy.

Many nations treated captured privateers as pirates so having a Letters of Marque and Reprisal was both a rewarding venture and a risky business. The Declaration of Paris, which was signed in 1856, discontinued the use of these letters among those nations signing the agreement.

Although the privateers did great service to our country during the War of 1812, they greatly hurt the recruiting efforts of the U.S. Navy. Seamen were more apt to join the crew of a privateer than a naval warship because of richer prize money. Privateers rarely engaged British warships but instead concentrated on the enemy's merchant vessels.

## Perry's Monument: How many are there?

Ask anyone where they can find Perry's Monument and they will probably tell you that it is located at Put-in-Bay, Ohio. They would be of course be referring to the Perry Peace Memorial Monument which was dedicated in 1915. This 352-foot monument is the most dominant feature in the western Lake Erie region and beneath the monument's rotunda are the remains of the six officers who were killed during the Battle of Lake Erie, three American and three British.

The original Perry's Monument was not a giant column of granite but a simple marble statue of Master Commandant Oliver Hazard Perry in a wind-swept poise. This statue was dedicated at Cleveland, Ohio, in 1860 having been sculptured by William Walcutt. It is said that this was the first public monument in the State of Ohio.

Over the years, the monument had four different location in Cleveland, two on Public Square, one at Wade Park and finally at Gordon Park. By 1929 the marble statue was showing its age and the Early Settlers Association of Cleveland commissioned Herman Matzen to cast two bronze statues from the original marble.

The first bronze was dedicated at Gordon Park on 14 June 1929. In 1951 it was relocated to a new location in the park and then in 1991 the bronze was moved to Fort Huntington Park in downtown Cleveland where it can be seen today.

The second bronze was given to the State of Rhode Island where it sits on the stairs on the south side of the Capitol Building in Providence. This statue has stayed put on the capitol steps unlike its roving twin in Cleveland.

The original marble statue was given to the City of Perrysburg, Ohio, where it was rededicated in 1937 overlooking the Maumee River in downtown Perrysburg. Perrysburg commissioned Lynn Hayes in 1997 to cast another bronze from the original marble. This bronze sits on a new pedestal where the original marble was located.

The City of Perrysburg donated the original marble statue to the International Peace Museum at Put-in-Bay where it is on display today inside the museum. The National Park Service maintains both the 352-foot monument and the museum.

Going back to the original question of how many Perry's Monuments are there? If you said four (not counting the granite monument) you are wrong!

The plaster cast which was used to cast the first two bronze statues sits in the entrance way of the Western Reserve Historical Society in Cleveland. This beautiful white statue is not on a pedestal so visitors can get a close up view of Perry. The location of the second plaster cast in not known by this author. It still may exist today.

Another statue of Perry sits in the Washington Square Park in Newport, Rhode Island, the home of this famous fighting sailor. Although it is not the same sculpture as first commissioned in Ohio, the statue does have a similar poise but with a more wind-swept look.

Besides the monuments, six communities in Ohio are named for Oliver Hazard Perry: Perry, Perrysburg, Perrysburg Heights, two Perrysvilles and Perryton. In addition one county and 25 townships are named after Perry.

Put-in-Bay, Ohio

Cleveland, Ohio

Providence,
Rhode Island

Western Reserve Historical
Society, Cleveland, Ohio

Perrysburg, Ohio

Newport, Rhode Island

## Ohio counties named after heroes of the War of 1812

Fourteen Ohio counties were named after men who were war heroes during the War of 1812 or who had served in important government positions, which contributed to the success of the American military.

Allen County was named for Colonel John L. Allen of the Kentucky militia who was killed during the Battle of the River Raisin in the Territory of Michigan.

Brown County was named for Brigadier General Jacob Brown of New York, U.S. Army. He led the American forces in the Battle of Chippewa, the Battle of Lundy's Lane and the Battle of Fort Erie, Upper Canada, in 1814. He was the commander of the U.S. Army between 1815 and 1828.

Harrison County was named for Major General William Henry Harrison of Ohio. He was the hero of the Battle of Tippecanoe, Territory of Indiana, and the Battle of the Thames River, Upper Canada. He became president of the United States.

Holmes County was named for Major Andrew Hunter Holmes of the Territory of Mississippi, U.S. Army, who was killed during the Battle of Fort Mackinaw, Territory of Michigan, on 4 August 1814. Major Holmes was a very popular officer under General Harrison.

Jackson County was named for Major General Andrew Jackson of Tennessee. He was the hero of the Battle of New Orleans. He became president of the United States.

Lawrence County was named for Captain James Lawrence of New Jersey, commander of the U.S. Frigate Chesapeake, who died on 4 June 1813 from wounds received during the engagement with the British frigate Shannon on 1 June 1813. He coined the phase "Don't Give Up the Ship."

Lucas County was named for Brigadier General Robert Lucas, Ohio Militia. Lucas was the governor of Ohio between 1832 and 1836 and the governor of the Territory of Iowa between 1838 and 1841.

Madison County was named for President James Madison, the fourth president of the United States and the commander-in-chief of the U.S. armed forces during the War of 1812.

Meigs County was named for Governor Return Jonathan Meigs, Junior, who was the governor of Ohio between 1810 and 1814. As governor he was the commander of the Ohio militia during the War of 1812.

Monroe County was named for James Monroe, Secretary of State (1811-1817), Secretary of War (27 September 1814-2 March 1815), and the fifth president of the United States. He was also the acting Secretary of War between December 1812 and January 1813.

Perry County was named for Master Commandant Oliver Hazard Perry of Rhode Island, U.S. Navy, who was the hero of the Battle of Lake Erie on 10 September 1813.

Pike County was named for Brigadier General Zebulon Montgomery Pike of New Jersey, U.S. Army. He was killed on 27 April 1813 during the Battle of York, Upper Canada. He discovered Pike's Peak in Colorado.

Shelby County was named for Governor Isaac Shelby of Kentucky who led the Kentucky militia during the Battle of the Thames River, Upper Canada, on 5 October 1813.

Wood County was named for Major Eleazer Derby Wood of New York, U.S. Army Engineers. He was the designer of Fort Meigs and Fort Stephenson. He served with distinction during the Battle of Lundy's Lane, Upper Canada. He was killed on 17 September 1814 during the Battle of Fort Erie, Upper Canada.

# United States Navy

The United States Navy was founded on 27 March 1794. On the eve of the War of 1812 the navy had in commission five frigates, three ships and seven brigs. A 'ship' was a three-mast vessel smaller than a frigate but larger than a two-mast brig.[18] Two of the vessels were on foreign service while one was operating on Lake Ontario. The rest of the vessels operated out of three ports along the eastern seaboard and one in the Gulf of Mexico. The navy also had five frigates in ordinary, that is, they were not in commission and they were laid up for repairs.

The navy also operated a 'brown water' force consisting of 165 gunboats stationed at twelve major ports along the east coast and the Gulf of Mexico. Of these crafts, 62 were in commission, 86 were laid up in ordinary, and seven were under repairs. The 'brown water' force was called the naval flotilla service, which was separate from the U.S. Flotilla Service.

Once war was declared, the U.S. Navy purchased and leased a number of vessels on the Great Lakes and along the sea coast in order to increase the size of the navy. The majority of these vessels were brigs, schooners and sloops.

By 1814 the navy was operating three 'de facto' fleets: one on the Atlantic Ocean, one on the Great Lakes and Lake Champlain, and the naval flotilla service.[19] The ocean-going navy had three ship-of-the-lines and three large frigates being built on the east coast and seven frigates, two corvettes, 11 sloops of war, five brigs and two schooners in commission.

Three squadrons operated on the Great Lakes and Lake Champlain. One squadron was headquartered at Erie, Pennsylvania, and this squadron operated on the upper Great Lakes. Sackets Harbor, New York, was the headquarters for the Lake Ontario Squadron while the final squadron operated on Lake Champlain. The Upper Great Lakes Squadron had four brigs, two ships, five schooners, and four sloops. The Lake Ontario Squadron had two ships, one brig, 12 schooners and a bomb ketch. The Lake Champlain Squadron had three sloops.

The naval flotilla service had 126 gunboats, 32 barges and 59 barges being built. It also had 11 armed vessels to support the flotilla service, which consisted on six schooners, three sloops, one bomb ketch, one cutter, and one pilot boat.

By the end of the war, the navy had three ships-on-the-lines, nine frigates, eight ships, ten brigs, three ketches, 13 schooners and three sloops operating on the high seas. The Lake Ontario Squadron had two ships-of-the-lines near completion, three frigates, two ships, four brigs, and a schooner. The Lake Erie Squadron had four brigs and two schooners while the Lake Champlain Squadron had two ships, two brigs and a schooner. The navy had two steam frigates being built on the east coast. The naval flotilla service had been disbanded and its vessels sold or laid up.

The U.S. Navy at the end of the War of 1812 was far larger and more powerful than it had ever been since its founding.[20] While the U.S. Army was reduced to its pre-war strength, the navy had been expanded from a defensive force to one that could better protect our maritime fleet anywhere in the world.

---

[18] *Increase of the Navy*, communicated to the House of Representatives, December 17, 1811, by the Secretary of the Navy, Twelfth Congress, First Session, number 87, American State Papers, Naval Affairs, pp. 247-252.

[19] *Condition of the Navy, and the Progress made in Providing Materials and Building Ships*, communicated to the Senate, March 18, 1814, by the Secretary of the Navy, Thirteenth Congress, Second Session, number 111, American State Papers, Naval Affairs, pp. 305-309.

[20] *Naval Force on the First of January 1816*, communicated to the Senate, January 5, 1816, by the Secretary of the Navy, Fourteenth Congress, First Session, number 133, American State Papers, Naval Affairs, pp. 379-380.

## Battle of the Maumee Rapids

On 14 November 1812, along the banks of the Maumee River in northwestern Ohio, during the War of 1812, the military forces of Brigadier General Edward W. Tupper of the Ohio Militia defeated a mounted Indian attack near the Maumee Rapids. Most historians and students of the War of 1812 dismiss this battle as a minor skirmish not worth writing about.

The question is: what is a battle and what is a skirmish? Historically, a battle is a military operation in which the two opposing forces plan their strategies and attack their opponent in military fashion on a pre-set battlefield. A skirmish is an unplanned military operation in which opposing forces meets unintentionally and begin to fight. Many skirmishes will lead into a full scale battle, such as what happened at Gettysburg during the American Civil War.

The second part of this scenario is what happens after the battle is over. Was that battle a strategic victory or a tactical victory? The Battles of Chippewa and Lundy's Lane were American tactical victories during the third year of the war but these two battles were also a strategic victory for the British, even though they lost both battles. The American objectives were to march down the Niagara River in Canada from Fort Erie and capture Fort George at the other end of the river, and then to proceed along Lake Ontario to Burlington Heights destroying the British army. Both armies were exhausted after the battles and the Americans withdrew back to Fort Erie. This was a strategic victory for the British since they stopped the Americans from obtaining their objectives.

The number of men involved in either a skirmish or a battle is irrelevant. There can be two men or two million men. What counts is the training and the discipline of the soldiers, and the abilities of a commander to plan an attack and to be able to adjust the plans to any condition.

General Tupper commanded the 2nd Ohio Volunteer Brigade consisting of a 1,000 militiamen who were ordered to Fort Detroit as six-month reinforcements for Brigadier General William Hull. On 16 August 1812 while the brigade was still being formed, General Hull surrendered his Army of the Northwest to the British at Fort Detroit.

Tupper's brigade was reassigned to Brigade General William Henry Harrison as Harrison was building the Army of the Northwest. The brigade was re-designated as the 1st Ohio Detachment and it was chosen as the central division of Harrison's new army. Colonel Charles Miller of Coshocton County became the commander of General Tupper's 2nd Regiment while Colonel Robert Safford of Gallia County became the commander of the 3rd Regiment. Major James Galloway of Xenia, Green County, was the commander of an independent battalion of mounted infantry and mounted spies, which was assigned to this brigade as the 1st Regiment. The 1st Regiment was from the southwestern counties of Ohio while the 2nd and 3rd Regiments were from the southeastern area of Ohio.

General Harrison's western division consisted of four militia regiments from Kentucky plus three companies of the 17th Regiment of U.S. Infantry and one company from the 19th Regiment of U.S. Infantry. This division was under the command of Brigadier General James Winchester. This division left Kentucky and headed for Fort Wayne, Territory of Indiana. From Fort Wayne the division headed down the Maumee River with hopes of meeting up with the other two divisions.

Brigadier General Simon Perkins led the 2nd Ohio Detachment which was organized from the counties in northeastern Ohio. General Harrison assigned this brigade as part of the eastern division of his army. The eastern division was made up of a militia brigade from Pennsylvania and another one from Virginia plus the brigade under General Perkins. There was no overall commander for the eastern division.

General Harrison's objectives were for the western division to come down the Maumee River from Fort Wayne to the rapids of the Maumee River while General Tupper would come up to the rapids using General Hull's road from Urbana, Ohio. General Perkins' brigade would travel to the rapids using the postal road between the rapids and Cleveland, Ohio. Brigadier General Richard Crooks of the 2nd Pennsylvania Brigade would come from Pittsburgh through Mansfield and then to the rapids while Brigadier General Joel Leftwich of the Western Virginia Brigade would travel from Point Pleasant, in now West Virginia, to Chillicothe and then to the rapids.

The three divisions were not able to meet at the rapids at the same time. General Winchester took it upon himself to advance to Fort Detroit after arriving at the rapids and he was defeated on 22 January 1813 at the Battle of the River Raisin. The eastern division arrived at the rapids before Winchester's division had left and they began the process of building Fort Meigs a few miles down the river from the rapids. Their enlistments expired before the fort was finished and they returned home.

General Tupper's brigade arrived at Fort McArthur in what is now Hardin County on 31 October 1812. Approximately, one-fourth of the detachment was unfit for duty due to sickness. The march from Urbana had taken its toll.

General Tupper sent Captain Hinkson and his spy company (rangers) to reconnoiter the rapids on 9 November 1812. The spies reported back that there were eighty mounted Indians warriors and fifty British soldiers, two gunboats, six batteauxs and a schooner at the rapids. The men returned with Captain Clark, a captured British officer (first name unknown). The purpose of the British raid was to carry off a quantity of corn near the rapids. The area of the rapids had been the home of seventy-five French-Canadian families who had left for Detroit after the war started in August.

On November 10th General Tupper prepared a detachment of 650 men and headed towards the rapids with each man having five days worth of rations. The other 350 men were either too sick to travel or were needed to guard the fort.

When General Tupper arrived at the Portage River on November 13th, 20 miles south of the rapids, he again sent forward a spy detachment to get an update on the activities of the enemy. This detail returned that night and reported that the British and Indians were encamped in a secure formation near the rapids on the western side of the river. The Indians were drinking heavily.

General Tupper's force then prepared to attack the enemy on the western side of the river the next morning. The force had a difficult time trying to cross the Maumee River. The current was swift and many men were washed down the river with their weapons. All of these men were saved.

The brigade withdrew to its camp near the rapids and then re-grouped with Colonel Safford commanding the left flank of the brigade while Colonel Miller commanded the right rank. Major Galloway's battalion was held in reserves.

A number of men from the left flank decided to leave formation and gather corn in a nearby field. A large force of mounted Indians attacked from behind the American lines and they killed four of the men. The left flank was ordered to attack this mounted force. In a twenty minute battle, the Indians were driven from the field as the left flank advanced in line formation The Indians were not able to break the American line.

A second wave of Indians tired to cross the river at the rapids and attack the right flank, which was also in line formation. The troops took minor causalities while the Indians suffered greatly and retreated back across the river. As the Indians were crossing the rapids, the troops fired upon them sending many Indians both killed and wounded, plus their horse, down the river. The Indians ended their assault.

During the assault by the Indians, the British did fire two of their cannons from their ship, which Lieutenant John Jackson claimed was the *Queen Charlotte*. The ship was a three-mast sloop-of-war carrying sixteen guns. It was not a schooner. The British had also marched their soldiers on the other side of the river with fifes and drums playing which was more of a scare tactic than an actual assault. The British soldiers were too few in number to try and engage the Americans.

Late in the evening General Tupper and his men headed back towards Fort McArthur, for their provisions and ammunition were almost exhausted. General Tupper's brigade remained at Fort McArthur for a short time and then they moved to the rapids as the eastern division arrived in the area. The four brigades (Tupper's, Perkins', Leftwich's and Crook's) began the construction of Fort Meigs but they were released from military duty in February 1813 when their tours of duty expired.

General Tupper, in this after action report, claimed that they were from 400 to 500 Indians engaged in the fighting. Overall, there were over a thousand men from both sides who participated in this battle.

The Ohio militia held their ground, remained in line formation, and followed the orders of their officers (accept for a few men). Tupper was able to change his tactics and to successfully defend against the Indians during a two sided attack. The causalities on the American side were four killed and a couple wounded. It is unknown how many Indians were killed and wounded in this engagement.

It was an important tactical victory for the Ohio militia, even though it was a 'minor' battle. Strategically, it was also an American victory since it forced the British and their Indian allies to withdraw back to Canada without having gather all of the food sources that they needed for their troops. Had the Ohioans not held their formation during this battle, this would have been a totally different article.

## Bibliography

Knopt, Richard C., *Return Jonathan Meigs, Jr. and the War of 1812*, (The Anthony Wayne Parkway Board, The Ohio State Museum: Columbus, Ohio 1961), Volume 2 of the Document Transcriptions of the War of 1812 in the Northwest.

Journal of Captain Samuel Black, Ohio Militia, *War of 1812 Reenactment Primary Source Journals and Diaries and Articles*, 12 February 2017, http://patcosta.com/war-of-1812-reenactment-primary-sources/.

*Journal of Nathan Newson*, transcribed by James Ohde, (Anthony Wayne Parkway Board: Columbus, OH 1957.

Palmer, T. H., *The Historical Register of the United States*, part II, volume II, second edition, (G. Palmer Publisher: Philadelphia, PA 1814).

"We lay there doing nothing:" John Jackson's recollection of the War of 1812, Indiana Magazine of History, volume LXXXVIII, number 2, June 1992, Indiana University, Bloomington, Indiana.

# Military Genealogy

# Finding Black soldiers in the early U.S. Army records

The American Black soldier was a rarity between the Revolutionary War and the American Civil War. It is said that approximately 5,000 free Blacks and slaves served in the Continental Army during the Revolution. George Washington needed men. The Blacks proved their worth, many fought with distinction, and many died for their country.

The U.S. Congress finally gave the president the authority to enlist Blacks as soldiers under the Militia Act of 1862.[1] Again, George Washington needed men. 209,145 Black men would serve in the United States Colored Troops and in the United States Navy during the Civil War. The Colored Troops would represent one-tenth of the total Union Army's strength during this war, while 40% of the Colored Troops would become causalities, either from the bullet or by disease.

Between these two wars, the Black man would be delegated to serve only as a company cook or a White officer's servant. The U.S. Navy would enlist Blacks as enlisted men on equal footing with the common sailor, however, the petty officer's and officer's ranks would still be off-limits to Blacks.

Cracks in this restriction appeared for a short time during the War of 1812. Without proper rules and regulations governing this issue, some commanders in the army did recruit Blacks although the total number of men was around 400. Louisiana did permit Blacks to serve in its state's militia, and approximately 800 men fought at the Battle of New Orleans in 1815.

When Congress passed the Militia Act of 1792,[2] it required that "every free able-bodied white male citizen" join his state militia. This act implied that non-whites could not participate in the militia, but it left open the possibility that non-whites could enlist in the U.S. Army. On 5 April 1799,[3] the War Department established the requirements for its recruiting service by stating, "natives of good character, are always to be preferred for soldiers. Foreigners of good reputation, for sobriety and honestly, may be enlisted; but Negroes, Mulattoes, or Indians are not to be enlisted."

The new ruling only limited Blacks from being soldiers but they could still serve as cooks and servants. In 1802[4] Congress helped the officers who hired private servants, or used their personal slaves as servants, by permitting the army to supply one ration (meal) per day for these individuals. Officers had always been allowed to take a soldier from within the officer's command to serve as a servant. These soldiers serving in this capacity were called "waiters."

Congress in 1812, when passing legislation to provide for the army, increased the support for officer's servants by stating, "that the officers who shall not take waiters from the line of the army, shall receive the pay, clothing and subsistence allowed to a private soldier." [5] This was a big boost for Black servants for

---

[1] Sander, George P., *The Statues at Large, Treaties and Proclamations of the United States of America*, Volume XII, (Little, Brown and Company: Boston, MA 1863), pp. 597-600, Thirty-seventh Congress, session II, chapter 201, *An act of amend the Act calling forth the Militia to execute the Laws of the Union, suppress Insurrections, and repel Invasions, approved February twenty-eight, seventeen hundred and ninety-five, and the Acts amendatory thereof, and for other Purposes*, section 12 and 13.

[2] Peters, Richard, *The Public Statues at Large of the United States of America*, Volume I, (Charles C. Little and James Brown: Boston, MA 1845), pp.271-274, Second Congress, 1st session, chapter 33, 8 May 1792, *An Act more effectually to provide for the National Defence by establishing an Uniform Militia throughout the United States*, section 1.

[3] *War Department, Rules and Regulations Respecting the Recruiting Service*, April 5, 1799, Article III – Natives of good character, are always to be preferred for soldiers. Foreigners of good reputation, for sobriety and honestly, may be enlisted; but Negroes, Mulattoes, or Indians are not be enlisted.

[4] Peters, *The Public Statues*, Volume II, (Charles C. Little and James Brown: Boston, MA 1845), pp.132-137, Seventh Congress, 1st session, chapter 9, 16 March 1802, *An Act fixing the Military Peace Establishment of the United States*, section 5.

[5] *Ibid.*, pp.784-785, Twelfth Congress, 1st session, chapter 137, 6 Jul 1812, *An Act making further Provision for the Army of the United States, and for other purposes*, section 5.

the U.S. Army would now feed, pay and cloth these men on an equal level with a private. Servants would now receive two rations (meals) per day, the same as privates.

Legislation in 1814 forbid officers from employing soldiers as servants and ruled that the names of the servants must now appear on the muster rolls for each corps.[6] Servants were now full members of the U.S. Army.

The Adjutant and Inspector General's Office of the U.S. Army issued a general order on 18 February 1820 stating that "no negro or mulatto will be received as a recruit of the army." This ended the Black man's participation as a soldier in the U.S. Army until the Civil War.

The army did record, to some extent, on its enlistment rosters whether the soldier was White or Black. The recorded physical description of a soldier also indicated race, but this was not always accurate.

The *Register of Enlistments in the U.S. Army, 1798-1914* [7] is an extract of personnel information from the land forces of the United States obtained from documents generated by the U.S. Army. These documents include enlistment records, recruiting reports, discharge records, commissioning records, muster rosters, inspection reports, morning reports, court-martial records, etc.

Finding Black soldiers in U.S. Army records needs to be divided into three time periods. The first period is between 1798 and 1812 and again between 1815 and 1862. The second period is during the War of 1812, between 1812 and 1815, while the final period starts in 1863.

The first period has a very small number of Blacks serving as cooks and servants simply because the size of the peace-time army was very small compared to the number of men serving during the War of 1812 and the Civil War. Black slaves serving as servants usually didn't have surnames, and nicknames were used as their given names, that is, Bob or Bill. In the comments section of the enlisted rosters, it may indicate that Bill was a Negro boy who was a servant for a particular officer while Bob was a Colored cook for a particular regiment. A researcher will find that the first period has very little, if any, genealogical value for these men.

The second period during the War of 1812 has great genealogical value because the army recorded the age and place of birth, date and place of enlistment, physical descriptions, and the comment section highlighted the career of a soldier. Two problems exist for this period: one, not all of the information is recorded for each soldier, and two, it is hard to spot a Mulatto, especially a very light skin Mulatto in many army records.

In the name column of the enlistment record there may be the words "Negro," "Black," "Colored," or "Mulatto" under the person's name. Like wise, these words may replace "private" in the rank column. "Colored" is often abbreviated as "Col'd." The comments section will also used these words but usually within sentences. Other terms that were used are "Negro boy" and "Blackman."

The army also captured the physical descriptions of a soldier by denoting the color of the eyes, the hair, and the complexion on the Register of Enlistments. During the 1800's, the term 'colored' denoted any one who wasn't White (Caucasian). In the records of the United States Colored Troops during the Civil War you will find Hawaiians, Mexicans and Filipinos, among other ethic groups, serving with the Blacks.

The color of the eyes will not denote race since all races had black eyes including Caucasians. Caucasians will also have red, blue, green, hazel, gray, and other colors. People of mixed races can also have eye color variants.

Most races will have either a dark brown or black hair while Caucasians will have many different hair colors. In the army records, 'curly' will also denote a Colored person.

---

[6] Peters, *The Public Statues*, Volume III, (Charles C. Little and James Brown: Boston, MA 1846), pp.113-116, Thirteenth Congress, 2nd session, chapter 37, 30 March 1814, *An Act for the better organizing, paying, and supplying the Army of the United States*, section 10.

[7] *Register of Enlistments in the U.S. Army, 1798-1914*; (National Archives Microfilm Publication M233, 81 rolls); Records of the Adjutant General's Office, 1780's-1917, Record Group 94; National Archives, Washington, D.C.

The recording on the complexion of a soldiers is the best indicator of race. The color black is rarely used for a Caucasian. The terms used for a Caucasian are light, dark, fair, ruddy, sallow, freckled, flesh, red, sandy or florid. Other terms do exist.

The terms used to describe Blacks or Mulattoes are black, brown, chestnut, or yellow. The last three terms usually denotes a Mulatto. The term 'yellow' is extremely interesting. In the 1800's, you were either White or Colored. "Yellow" was used to denote a mixed race person who wasn't fully White or fully Black. It did not mean that the person was Asian.

Using the three physical characteristics of a soldier may indicate either a Black or Mulatto soldier. These are:

| Eyes | Hair | Complexion |
| --- | --- | --- |
| Black | Black | Black |
| Black | Curly | Black |
| Black | Black | Yellow |
| Black | Curly | Yellow |
| Brown | Brown | Brown |

The Mexican-American War presents a real problem in identifying Blacks who served during this conflict. Blacks were still not permitted to serve as soldiers, only as servants and cooks. Light-skinned Mulattoes are know to have enlisted in the army but they are extremely hard to identify in the enlistment reports.

Since this conflict was our first 'true' overseas war, the army requested that officers leave their personal servants and their company cooks behind in the States, for they could hire Mexicans very cheaply to do these same tasks. Most of the American soldiers embarked for Mexico at New Orleans and the army did not want to pay to transport cooks and servants to Mexico and back.

You almost have to know that your ancestor served in this war as a cook or servant, or possibly as a soldier, first before finding them in army records. Family history and traditions are extremely important in order to do Black research during the Mexican-American War.

The third period of Black military research begins in 1863 and it won't end until the mid-1950's when the military was fully integrated. This period witnesses the formation of Black regiments serving in the U.S. Army, first as Volunteer regiments and then as regular army regiments.

These regiments, at first, were officered by Whites with Black enlisted men. During the Philippine-American War (1899-1902), the regimental staffs had White officers while the companies had Black officers and enlisted man. World War I opened the staff positions to Black officers. This article will only deal with the Civil War regiments.

The real problem with Black military research during the Civil War is not identifying Blacks but identifying Caucasians and other ethic groups who served in Black regiments. Although these regiments were commanded by White officers, 120 Blacks did receive volunteer commissions and they did served in Black regiments. *The Military Order of the Loyal Legion of the United States* (MOLLUS) maintains a webpage listing these Black officers and the regiments that they served in.[8]

Another surprise in Black military research during the Civil War period is that the army required that senior sergeants, that is, the Sergeant Majors, the Quartermaster Sergeants, and the Commissary Sergeants, to be able to read, write and do arithmetic (the 3 R's). This was not a roadblock for promotion. These three ranks had reports to fill out, orders to read, food to purchase, and clothing and tents to order, among other duties. To some extent, knowing the 3 R's would help each company's First Sergeants better perform their duties.

Many Black regiments, particularly in the South, where it was illegal to educate slaves, had White senior enlisted personnel until Blacks could be found to replace them. Once these White sergeants were

---

[8] *Military Order of the Loyal Legion of the United States*, United States Commissioned Officers of African-American Descent who served during the Civil War, http://suvcw.org/mollus/usctofficers.htm

replaced, many would be promoted to the officer's ranks. Many recruiters from the South operated in northern states throughout the war seeking educated Blacks for their senior sergeant's positions.

There were a number of Blacks who served in White regiments during the war as soldiers and not as non-combatants. William Mulligan, John Muncy, Montgomery Muncy and William Muncy were all cavalrymen in Company L of the 1st Ohio Volunteer Cavalry. They were all transferred to the 9th United States Colored Heavy Artillery Regiment in January 1865. It is not known if these were voluntary or involuntary transfers. Most Black soldiers in White regiments would transfer to the United States Colored Troops. Many light-skinned Mulattoes served in White regiments and stayed in these regiments.

The *Official Roster of the Soldiers of the State of Ohio in the War of the Rebellion* does list the colored cooks in Ohio's White regiments.[9] These men are listed as "Cooks" or "Colored Under Cooks" after the listing of privates in the companies of some of the regiments. Only a few regiments had Colored cooks and the total number is around 400. White cooks were privates from their companies who were detailed to serve as cooks.

Although, it can to difficult to research Black soldiers prior to 1863 in the U.S. Army, many Blacks who served during the War of 1812 did receive pensions and land bounties. Understanding how the army kept its descriptive records and enlistment reports can greatly increase the success in researching Black soldiers in the early U.S. Army records.

---

[9] *Official Roster of the Soldiers of the State of Ohio in the War of the Rebellion, 1861-1866*, 12 volumes, (Werner Company. Akron. Ohio. 1886-95).

# The Army's Register of Enlistments pertaining to the War of 1812

The book, *Register of Enlistments in the U.S. Army, 1798-1914,* is an extract of personnel information for the land forces of the United States obtained from documents generated by the U.S. Army. This book can be found at the National Archives in Washington, D.C. and on-line through Ancestry.com.

The first section of this book covers the period from 1798 to 17 May 1815. This section of the book is also called the *Records of the Men Enlisted in the U.S. Army Prior to the Peace Establishment, May 17, 1815.* The majority of the 7,360 pages in this section include the enlistments and commissions of men who served during the War of 1812.

The Army branches included in this section are the infantry, light dragoons, artillery, rifles and rangers. Men who served in the various army departments are also listed. These department are the adjutant general's, inspector general's, judge advocate general's, quartermaster's, subsistence, medical, pay, corps of engineers, topographical engineers, and ordnance.

Officers and enlisted men who served on the staffs of the regiments, brigades and division plus the auxiliary units (U.S. Volunteers, Sea Fencibles, Canadian Volunteers and West Point cadets) are also listed. There are entries for militiamen and British prisoners of war if they were mentioned in any army document. Finally, the men from the U.S. Marines can also be found in this register. If your ancestor served in the War of 1812, it may be beneficial to search these pages even though your ancestor did not serve in the U.S. Army.

The U.S. Army required a number of different types of reports to be completed in a timely manner. Certain reports were required daily, weekly, monthly, quarterly, or semi-annually. Each regiment maintained their own 'recruiting service' and as such, produced weekly recruiting reports. These reports were consolidated by each recruiting district into monthly and quarterly reports. Besides the recruiting reports, there were initial clothing and equipment reports needed to be complete for the new soldiers.

A Muster Roll containing the names and ranks of all of the men in each company was required every two months. These reports were completed on the last day of February, April, June, August, October and December. Another type of muster roll was required each month showing only the number of men by each rank who were fit for duty. This report also included the names of the men who were not fit for duty, where they were located and why they were not fit for duty. In most cases, these men were listed by their surnames and the first initial of their given name.

An inspection report was required quarterly. This report listed each man in a company plus what type of small arms that man had been issued. Both the muster roll and the inspection report were required as a semi-annual report covering a six-month period ending each June 30th and December 31st.

While in camp, regiments were required to produce Morning Reports. These reports listed the number of men by rank who were fit for duty that day, who were on sick call, who were in hospitals, who recently had died, or who was 'on command.' A soldier 'on command' was performing duties away from his company.

Each company maintained a Descriptive Roll listing every man in the company, his age, height, weight, color of eyes and complexion, where born, where enlisted, the name of the officer who enlisted them, enlistment date, length of enlistment, and bonus amounts that were paid and what was due. These reports could also include a list of clothing and other supplies that were issued to the soldiers.

A Clothing Report was required when new uniforms or replacement uniforms, new equipment or replacement equipment were issued to the men. Paydays in the army occurred every two months at the end of odd numbered months. Paydays produced payroll reports.

The last required muster was the Muster Out Report, which ended the military service of an individual. Normally all of the men within a regiment who were being discharged were organized into a 'muster out' detachment. The commanders of these 'muster out' detachments were normally officers who were also being discharged, usually lieutenants or ensigns.

Company, regiments and brigades were required to maintain an Orders (or Orderly) Book and a Record Book. All up and down the chain of command clerks copied reports and letters and then posted them in these books. Depots, forts, recruiting facilities, hospitals and other military installations also were required to maintain an orderly book and a record book. General orders and special orders were written down, copied many times and distributed to all concerned parties.

Regiments were also required to keep copies of the Description Rolls, all correspondences, and all monthly returns. Other types of reports include a soldier's transfer papers from one company or regiment to another, court martial proceedings, prisoner of war listings, discharged certificates and certificates of death.

As stated, the *Register of Enlistments in the U.S. Army, 1798-1914* contains personnel information extracted from army reports. This information includes the name and rank of each soldier or marine, his regiment, his company commander's name, his regimental commander's name, a physical description of the soldier plus his peacetime occupation, where he was born (county and state, or country), where he enlisted and the period of enlistment, and any additional remarks. All of the columns may not be filled in.

**List of abbreviations found in the *Register*:**

MoRet or MRet - Monthly Returns
MR - Morning Reports or Muster Rolls
RR - Recruiting Returns
DR - Descriptive Rolls
GO - General Orders
SO – Special Orders
IR - Inspection Returns
CoBook - Company book
SAMR – Semi-Annual Muster Rolls
SAIR – Semi-Annual Inspection Returns
PR - Payroll Reports
OB - Orders Book or Orderly Book

## Bibliography

*Register of Enlistments in the U.S. Army, 1798-1914*; (National Archives Microfilm Publication M233, 81 rolls); Records of the Adjutant General's Office, 1780's-1917, Record Group 94; National Archives, Washington, D.C.

Adjutant and Inspector General's Office, *Military Laws and Rules and Regulations for the Army of the United States*, September 1816, (E. De Krafft, Printer: Washington, DC).

# Thirty-seven 1812 veterans are interred in the Dayton National Cemetery

Thirty-seven veterans of the War of 1812 are interred in the Dayton National Cemetery in Dayton, Ohio. Each of these men were residents of the Central Branch of the National Asylum for Disabled Volunteer Soldiers at the time of their deaths. The Asylum is commonly called the Dayton Soldiers Home. It is now the Dayton Veterans Administration Medical Center.

The Ohio Chapter of the National Society United States Daughters 1812 dedicated a bronze plaque which was affixed to a tall boulder on 10 September 1936 in memory of these soldiers. The inscription on the plaque states that there were thirty-three veterans buried in this cemetery.

Memorial to the thirty-three soldiers of the War of 1812 buried in this cemetery
Honoring Josephine C. Diefenbach, state president 1935-1937
Erected by the Ohio Society United States Daughters of 1812
On the Anniversary of Perry's Victory – September 10, 1936

The names of the name were extracted from the *U.S., Burial Registers, Military Posts and National Cemeteries, 1862-1960* for the Dayton National Cemetery while the service of each of the men was found in the *U.S. National Homes for Disabled Volunteer Soldiers, 1866-1938* records. Each of these ledgers can be found on Ancestry.com.

Ohio Society United States Daughters of 1812
Memorial Stone – Dayton National Cemetery

| Name | Service | Date of Death | Gravesite - Section-Row-Number |
|---|---|---|---|
| Anderson, Ira | Kentucky troops | 26 Sep 1876 | A-4-56 |
| Andross, John P. | War of 1812 | 26 May 1880 | B-8-6 |
| Bailey, Joshua | Ohio militia | 28 Aug 1881 | C-7-12 |
| Barber, James | New York militia | 24 Dec 1887 | G-7-6 |
| Brent, Charles | Virginia militia | 18 Jan 1881 | B-17-32 |
| Brown, Thomas | War of 1812 | 27 Sep 1887 | G-3-6 |
| Conyers, John | Kentucky militia | 13 Nov 1884 | E-10-7 |
| Crosby, Jeremiah T. | War of 1812 | 9 Aug 1883 | D-17-5 |
| Daniels, George W. | 34th US Infantry | 10 Jun 1879 | A-18-41 |
| Davis, Justice | War of 1812 | 23 Feb 1885 | D-6-9 |
| Eutrican, James | New York militia | 8 Jun 1881 | C-?-6 |
| Figley, Simon | Ohio militia | 24 May 1883 | D-12-7 |
| Fisk, Jonathan | 1st US Rifles | 5 Jun 1884 | E-5-8 |
| Goodrich, Henry | 27th US Infantry | 6 Mar 1883 | D-10-11 |
| Hoagland, Enoch | War of 1812 | 30 Nov 1878 | A-13-12 |
| Holden, William G. | 45th US Infantry | 20 Jun 1882 | C-12-19 |
| also 55th Illinois Volunteer Infantry (Civil War) | | | |
| Hutchins, James | Quartermaster Dept. | 10 Apr 1873 | A-8-68 |
| Jones, Elias | New York militia | 24 Jan 1883 | D-10-4 |
| Jones, Evan | 22nd US Infantry | 9 Nov 1877 | A-1-31 |
| Leroy, Benjamin F. | Vermont militia | 6 Oct 1878 | A-14-25 |
| also 11th US Infantry (War of 1812) | | | |
| Maddox, William | Maryland militia | 16 Jan 1878 | A-1-9 |
| McClaren, Charles | Virginia militia | 9 Apr 1879 | A-17-17 |
| McDonald, William | Virginia militia | 19 Jan 1879 | A-18-6 |
| McGuire, Peter | 23rd US Infantry | 11 Dec 1877 | A-2-21 |
| Miller, Peter W. | New York militia | 27 Dec 1889 | H-12-11 |
| Parrish, Abraham L. | Kentucky militia | 3 Mar 1878 | A-6-41 |
| Peak, Elijah | New York militia | 10 Jan 1878 | A-2-12 |
| Rowe, Frederick H. | New York militia | 3 Feb 1883 | D-9-6 |
| Salyers, John W. | Ohio militia | 4 Jan 1875 | A-5-46 |
| Smith, Charles | 7th Infantry | 11 Oct 1887 | C-9-11 |
| served 45 years in the US Army and 6 years in the US Navy Ohio militia | | | |
| Smith, Isaac | New York militia | 11 Apr 1883 | D-9-18 |
| Son, Jacob | Pennsylvania militia | 24 Jul 1880 | B-9-12 |
| Spunagle, Samuel | 36th US Infantry | 12 Jan 1881 | B-17-29 |
| Walton, Mathew | New York militia | 10 Aug 1874 | A-5-63 |
| Warnick, Robert | 12th US Infantry | 29 Nov 1881 | C-12-5 |
| Whitney, Nehemiah | War of 1812 | 12 Sep 1869 | A-12-27 |
| Woodman, Enoch | War of 1812 | 9 Aug 1878 | A-13-29 |
| also 1st Michigan Volunteer Engineers (Civil War) | | | |

# The American Fencibles:
# The Forgotten Side of the Early Military

The term 'fencibles' is usually not associated with the American military but fencibles units were raised between the end of the Revolutionary War and the end of the War of 1812. This term is found in the British military system beginning in the early 1700's through the War of 1812 when describing a certain type of military unit.

Simply, fencibles were locally raised military units used as garrison troops in order to free combat troops for oversea duties. Fencibles could be raised from existing militia regiments or from men who have been pensioned off from the military. Normally, these troops were not as fit as the line troops, usually, because they were older men who did not have the stamina to keep up with the younger men. But, they were ideally suited to serve as guards in the forts on the home front. Most of the fencibles were raised once hostilities started and then disbanded after the signing of a treaty of peace. These fencibles units were a part of the regular British army, subject to the same rules and regulations.

In the British colonies, fencibles were raised locally to defend the colonies. Upper and Lower Canada and the Maritime Provinces of Canada raised many fencibles regiments and battalions, which served commendably during the War of 1812. These fencibles were combat units. In most cases, the term 'fencibles' is used as part of the unit's numeric designation and name.

The American fencibles units were raised during three periods in the early days of the republic. Under the Articles of Confederation, the army operated with restrictions to the length of service for each regiment and to the location within the country where each regiment could operate. During the early Indian wars after 1787 the army again operated fencibles under the current Constitution. The final phase occurred during the War of 1812 when fencibles were raised to replace regular army regiments so that they could be used on the frontier.

Only one type of American military organization used the term 'fencibles' in their military designation and that was the ten sea fencibles companies, which were organized in 1814. In order to distinguish the other fencibles from the normal military units you need to read the acts of Congress, which created those units.

## The fencibles under the Articles of Confederation

The bulk of the Continental Army was disbanded on 2 June 1784 upon a resolution adopted by the Congress, which was operating under the Articles of Confederation. What remained of this once mighty army, which served so faithfully during the Revolutionary War, was limited to 25 privates at Fort Pitt, Pennsylvania and 55 privates at West Point, New York, with an appropriate number of company-grade officers. The other soldiers were discharged with two months additional pay. This final Continental Army company, under the command of Captain John Doughty, would remain intact until it became a part of the Battalion of Artillery which was created on 29 September 1789.

The next day Congress created a regiment of 700 militiamen, which would serve under federal control in the Ohio River valley for up to a one-year period. The purpose of this regiment was to take possession of the western forts that were being evacuated by the British Army after the end of the Revolutionary War, to protect the legal settlers in the northwestern area of the United States, and to guard the public stores at Fort Pitt. This regiment was authorized to have eight infantry companies and two artillery companies.

This regiment was called the American Regiment. It was also called the 1st American Regiment or Harmar's Regiment after the name of its commanding officer, Lieutenant Colonel Commandant Joshua Harmar.

The Founding Fathers were very skeptical of a standing army and for the next 30 years Congress would continue to raise regiments for a given period of time, which would serve only in a given location, thus all of the early army regiments were "fencibles." This new regiment would be raised by the states but armed, equipped, and clothed by the federal government. Officers were appointed and commissioned by the President of the Congress.

On 1 April 1785 the regiment was continued for three more years with the same restrictions as listed under the resolution of 3 June 1784. A second regiment was authorized for three years on 20 October 1786. Since the enlistments of the majority of the troops were expiring in the coming year, Congress on 3 October 1787 merged the two regiments into a single regiment of 700 men to serve for three years in the Northwest Territory. This regiment was tasked to protect the settlers on public lands, to facilitate the surveying and selling of government lands, and to remove squatters from government lands. The regiment was thinly spread out among three forts in western Pennsylvania, three forts in what is now Ohio, and a single fort in what is now Indiana.

## The fencibles of the early Indian Wars

With the formation of the new government under the current Constitution, Congress on 29 September 1789 passed an act, which would disband the existing militia army after the end of the second term of the First Congress. Like its predecessor, the Continental Army, the militia army formed under the Articles of Confederation would be dismissed. The army in 1789 consisted of an infantry regiment made up of three infantry battalions of four companies each and an artillery battalion of four companies.

The regular army was established on 30 April 1790 when Congress passed an act authorizing the raising of a regiment to serve on the frontier for a period of up to three years. The regiment raised under the Articles of Confederation consisted of militiamen while this new regiment was now opened to all men with enlistments controlled by the federal government.

Those soldiers from the American Regiment who wished to continue to serve in the army had to re-enlist into the Regiment of Infantry. The American Regiment became the Regiment of Infantry on 30 April 1790 and then it was renamed the 1st Regiment of Infantry on 3 March 1791.

The Act of 3 March 1791 was an extension to the previous act, which established the United States Army as a professional military organization. This act doubled the size of the army by authorizing a second regiment to be organized under the same conditions of the original regiment. Under this same act the President was also given the authority to raise levies for a period of not more than 6 months consisting of 2,000 enlisted men plus officers. The President appointed and commissioned the levy officers.

The levies were not drafted or conscripted men as depicted in many writings. These men were volunteers who enlisted to serve for six months with the army instead of the three-year enlistment required for service in the regular regiments. They could only be used to protect the inhabitants of the frontiers of the United States from hostile Indians. Each of the two levy regiments were made up of three infantry battalions of four companies each. Many of the men in the levies re-enlisted in the regular regiments when the levy regiments were disbanded in early 1792.

Under the Act of 5 March 1792, three addition infantry regiments were authorized for the duration of the Indian wars in the northwest. One of the regiments was to be organized with two infantry battalions and a dragoon squadron of four troops.

The first non-fencible unit that was raised for the U.S. Army was the Corps of Artillerists and Engineers authorized under the Act of 9 May 1794. This corps could serve for three years or less in the field, on the frontiers or in fortifications along the seacoast, basically, anywhere within the boundary of the United States. The corps was made up of four battalions, each with four companies.

Between 3 March 1795 and 16 March 1802 the federal government continued to set a time limit on when a regiments could serve but the government no longer restricted the regiments from serving in a particular area of the country.

## The fencibles during the War of 1812

The Act of 16 March 1802 created the Military Peace Establishment, which authorized the U.S. Army to have a Corps of Engineers, an artillery regiment, and two infantry regiments. These units were not restricted to a length of service or to a certain location within the United States. This date marks the true beginnings of the professional standing army in the United States.

Also on this date the Military Academy at West Point was established. This institute had its beginnings as an engineering school training cadets to become engineer and artillery officers. After the War of 1812 the academy would broaden its curriculum to include infantry and cavalry cadets.

From 16 March 1802 until 3 March 1815 all corps, regiments, battalions and companies that were authorized by Congress had an operating restriction of five years, five years or during the war or of one year before they were disbanded. Some of these units had restrictions as to where they could serve, thus, these units were fencibles.

Fencible regiments were organized from existing infantry regiments. All of the ranger companies and all of the sea fencibles companies raised during the War of 1812 were fencible companies. The final fencible organization was actually an independent service raised to supplement the U.S. Navy. This service was called the U.S. Flotilla Service.

## Infantry Regiments

On 29 January 1813 Congress created 20 infantry regiments to serve in the War of 1812 but only the 26th through 44th Regiments of Infantry were raised for a term of one year with one-year enlistments. Under the act of 25 February 1813 the final regiment was converted to ten ranger companies for one-year service.

Under the Act of 5 July 1813 the 40th through the 44th Regiments of Infantry were converted into fencible regiments with 'during the war' enlistments. These regiments could only serve in the defense of the seaboard of the United States and they could not be sent to serve in other parts of the country or in foreign territories. This restriction was used to help recruit men who otherwise probably would not have joined the army.

The 41st through the 43rd Regiments of Infantry were actually artillery regiments assigned to forts along the eastern seaboard. They had incorporate state sea fencibles and other state artillery companies when they were formed. After the war these three regiments were merged into the Corps of Artillery. The remaining two fencible regiments merged into new infantry regiments that were created after the war.

## The Rangers

Six companies of mounted rangers were authorized under the act of 2 January 1812 for one year and by the act of 1 July 1812 an additional company was authorized under the same provisions and restrictions. The rangers were raised to combat an actual or threatened invasion of any state or territory by any Indian tribes or tribes. They were raised for duty in the old Northwest Territory replacing the 4th Regiment of Infantry after this regiment had been reassigned to New England.

The Act of 25 February 1813 authorized ten additional ranger companies for one year in lieu of one of the regiments of infantry authorized by the Act of 29 January 1813. The original seven ranger companies were continued for another year of service by the Act of 24 July 1813. By the Act of 24 February 1814 the ten new companies were retained in service for another year. All of the ranger companies were disbanded on 15 June 1815.

## The Sea Fencibles

The ten companies of Sea Fencibles were created under the Act of 26 July 1813 to serve for a one-year period. These were coastal defense companies of local men who guarded the ports and harbors along the seacoast and who were not subject to overseas duty.

The Sea Fencibles were unique since they were made up of army officers and naval enlisted men. The Sea Fencibles were an extremely distinctive fighting force since they could act as infantry, man artillery batteries and operate the gunboats in the harbors of our major ports. Section five of the Act of 26 July 1813 authorized the continuation of the Sea Fencibles until the end of the war. The sea fencibles were disbanded on 15 June 1815.

## U.S. Flotilla Service

Congress created the U.S. Flotilla Service on 16 April 1814 as a separate military service. This service was not a part of the navy but it was still under the control of the Secretary of the Navy. Four captains and twelve lieutenant positions were authorized for this service with the same relative rank and authority as the same grade in the navy.

The gunboats and barges of the Baltimore Flotilla, the Potomac Flotilla, and the New York Flotilla, operated by the U.S. Navy, were turned over to the U.S. Flotilla Service. Captain Joshua Barney took command of the Chesapeake Bay unit made up of the former Baltimore and Potomac flotillas while Captain Jacob Lewis commanded the New York unit.

In a letter to the U.S. Senate outlining the condition of the navy and the progress of the naval construction for 1814, the Secretary of the Navy William Jones said that the purpose of the U.S. Flotilla Service was to replace the officers and men in the naval flotilla service with local family men who wanted to serve in the navy but who did not want sea duty. Experienced naval officers were badly needed for the new ships being built on the east coast and for our naval squadrons on the Great Lakes. Most officers and men in the naval flotilla service wanted sea duty.

On 27 February 1815 the U.S. Flotilla Service was disbanded and its barges were either sold or laid up. All of the men in the service were discharged and given four months extra pay.

## Conclusion

The fencibles created under the Articles of Confederation and during the first years of the current Constitution were created simply because Congress feared a standing army. Under the Articles of Confederation the army could only operate in the old Northwest Territory while in the early Constitution era the army could operate anywhere on the frontier. Most of the army served in the old Northwest Territory but there was a detachment of infantry and artillery companies stationed in Georgia.

During the War of 1812 fencibles were raised simply to replace regular army regiments so that they could be redeployed to a combat area. Fencibles were used to protect the seaboard of the United States and to operate against warring Indian tribes in the old Northwest Territory.

## Bibliography

Wright, Robert K., Jr., and Morris J. MacGregor, Jr., *Soldier-Statesmen of the Constitution*, (Washington, D.C.: Center of Military History, United States Army, 1987), Selected Documents: The Revolutionary Years, pages 193, 209-211.

Heitman, Francis B., *Historical Register and Dictionary of the United States Army From Its Organization, September 29, 1789, to March 2, 1903*, Volume I, (Genealogical Publishing Company, Baltimore, Maryland: 1994), page 50, Artillery, Battalion of Artillery.

*Public Statues at Large of the United States of America*, volume I, (Boston, Massachusetts: Charles C. Little and James Brown, 1845), First Congress, session I, chapter XXV, pp. 95-96, 29 September 1789, an act to recognize and adapt to the Constitution of the United States the establishment of the troops raised under the resolve of the United States in Congress assembled, and for other purposes therein mentioned.

*Ibid*, session II, chapter X, pp. 119-121, 30 April 1790, an act for regulating the Military Establishment of the United States.

*Ibid*, session III, chapter XXVIII, pp. 222-224, 3 March 1791, an act for raising and adding another regiment to the Military Establishment of the United States, and for making farther provision for the protection of the frontiers.

*Ibid*, Third Congress, session II, chapter XLIV, pp. 430-432, 3 March 1795, an act for continuing and regulating the military establishment of the United States, and for repealing sundry acts heretofore passed on that subject.

*Ibid*, Seventh Congress, session I, chapter IX, pp. 132-137, 16 March 1802, an act fixing the military peace establishment of the United States.

*Ibid*, Twelfth Congress, session II, chapter XVI, pp. 794-796, 29 January 1813, an act in addition to the act entitled "an act to raise additional military force," and for other purposes.

*Ibid*, Twelfth Congress, session II, chapter XXXI, page 804, 25 February 1813, an act to raise ten additional companies of Rangers.

*Ibid*, volume III, Thirteenth Congress, session I, chapter IV, page 3, 5 July 1813, an act to amend the act in addition to the act, entitled "An act to raise additional military force, and for other purposes."

*Ibid*, Twelfth Congress, session 1, chapter XI, page 670, 2 January 1812, an act authorizing the President of the United States to raise certain companies of Rangers for the protection of the frontier of the United States.

*Ibid*, Twelfth Congress, session 1, chapter CXIX, pp. 774-775, 1 July 1812, an act supplementary to "an act authorizing the President of the United States to raise certain companies of Rangers for the protection of the frontier of the United States."

*Public Statues at Large of the United States of America*, volume II, (Boston, Massachusetts: Charles C. Little and James Brown, 1845), Twelfth Congress, session II, chapter XXXI, page 804, 25 February 1813, an act to raise ten additional companies of Rangers.

*Ibid*, volume III, Thirteenth Congress, session I, chapter XXIII, pp. 39-40, 24 July 1813, an act to continue in force for a limited time, certain acts authorizing corps of rangers for the protection of the frontier of the United States and making appropriations for the same.

*Ibid*, Thirteenth Congress, session II, chapter XV, page 98, 24 February 1814, an act to continue in force an act to raise ten additional companies of rangers.

*Ibid*, volume III, Thirteenth Congress, session I, chapter XXVII, pp. 47-48, 26 July 1813, an act to authorize the raising a corps of sea fencibles.

Ibid, Thirteenth Congress, session II, chapter 59, 16 April 1814, page 125, an act authorizing the appointment of certain officers for the flotilla service.

Dudley, William S., *The Naval War of 1812, A Documentary History*, three volumes, Washington, DC: Naval Historical Center, Department of the Navy.

*Public Statues at Large of the United States of America*, volume II, (Boston, Massachusetts: Charles C. Little and James Brown, 1845), Thirteenth Congress, session III, chapter 62, 27 February 1815, pages 217-218, an act to repeal certain acts concerning the flotilla service, and for other purposes.

# The U.S. Voluntary Corps

An often-overlooked area of the War of 1812 is the role in which the United States Voluntary Corps played in defending our nation. This corps was made up of militiamen who volunteered to serve with the U.S. Army for one year. They were under federal control and the records of these regiments and companies that were formed for this corps are usually not found with the various state and territorial militia records.

This corps gave extraordinary service in defending our country. Many of these companies and regiments fought shoulder-to-shoulder with the regular army and many times out fought both the army and the traditional militia troops.

The story of the voluntary corps actually begins 30 years earlier when Congress created the Regiment of U.S. Infantry on 3 June 1784. This regiment was made up of militia companies that were activated for a one-year tour of duty and who were clothed, armed and equipped by the federal government for duty in the Northwest Territory.

Congress maintained the regiment's existence through yearly resolutions until 29 September 1789 when this regiment became a permanent establishment in the army. During the Indian wars of the 1790's Congress also created three 6-month regiments again made up of militiamen who served in the northwest on federal duty. These regiments were called the 1st through 3rd U.S. Levies and these regiments were also clothed, armed and equipped by the federal government.

The Voluntary Corps during the War of 1812 was supposed to have supplemented the army until the army had built itself up to full strength. Very few army regiments were ever at full strength so the corps was maintained throughout the war.

The main difference between the volunteers, as the voluntary corps was called, and the regular militia is that the federal government from the established state and territorial militia organizations raised the volunteers. The traditional militia used by the federal government, on the other hand, was raised and equipped by a state or territory and then sent to serve with the U.S. Army on 'tours of duty' lasting up to 6-months.

Few of the traditional militia units were used in any combat situation during the war. The noted exception is the militia from New York, Pennsylvania, Kentucky, Maryland and Tennessee. Most of these militia units served for less than six months, usually for a specific task, and then they were released from duty once that task was completed.

Only a few of the thousands of militiamen who were called from their homes to protect this country ever saw combat or were even near an area of potential danger. Most of these men served in blockhouses or forts on guard duty, others transported supplies from one post to another while others served as messengers delivering dispatches. Many militia regiments were called to duty only to find that they had been tasked to build military roads and facilities.

Many of the militia companies, which served in the Voluntary Corps, were elite units that had been formed in the larger villages and cities before the war. These units were uniformed along European standards, well equipped and well armed. Many of these companies were better trained than their regular army counterparts. These companies had colorful names, such as, the "Pittsburgh Blues" or the "Zanesville Light Horse" or the "Baltimore Volunteers". When serving a "tour of duty" they had no trouble serving for a year and most of these companies were attached directly to regular army regiments or brigades.

## The Corps

The United States Congress created the U. S. Voluntary Corps under the Act of 6 February 1812. This corps was made up of militiamen who would serve a one-year enlistment with the Army of the United States. They were the first militia units to be called to duty before the war started.

The President of the United States (President) had the power to accept into military service up to 50,000 militiamen in companies of artillery, cavalry and infantry. Those militiamen in the cavalry had to supply their own horses while all members of the corps had to provide their own uniforms. The federal government would arm and equipped all of these volunteers.

Besides accepting militia companies into the Voluntary Corps, the president could also accept militia battalions, regiments, brigades and divisions. These militia units would be raised by their state or territory and the officers would be commissioned through their state or territorial militia laws.

The President could call to duty the companies that were designated to be apart of the Voluntary Corps within a two-year period of time and they would serve for 12 months starting when they had arrived at their place of rendezvous.

These soldiers were covered by the same rules and regulations as the regular army and these men were entitled to the same benefits as regular soldiers except for land bounties and yearly clothing allowance. Non-commissioned officers and soldiers were entitled to receive a one-time allowance for clothing equal to the clothing allowance for the regular army.

The President had the authority to organization the volunteer companies into battalions, squadrons, regiments, brigades and division as needed. If a member of the corps had lost his horse or the horse was injured and if his personally supplied equipment was lost or damage, he could be reimbursed for the value of the item or the horse.

The volunteers were entitled to a pension if they were wounded or injured in the line of duty. If the volunteer had died or was killed in the line of duty then his heirs would be entitled to 160 acres of land or half pay for five years.

Any non-commissioned officer or soldier who was discharged after completing only one month of service was entitled to keep his musket, bayonet and other personal equipment if he served with the infantry or the artillery. Those serving with the cavalry were allowed to keep their sabers and pistols. In 1812 one million dollars were appropriated for the expenses of the Voluntary Corps.

On 6 July 1812 Congress passed a supplementary act giving the President more power in controlling the Voluntary Corps.[10] The President could appoint and commission officers in the Voluntary Corps as long as these men had signed an enrollment for service in the corps. All officers in the corps would be commissioned by the President and not by the various state or territorial governments.

The President could now organized the corps into battalions, squadrons, regiments, brigades and divisions and to appoint (with the consent of the U.S. Senate) all generals, field officers and staff officers who would have the same pay and benefits of regular army officers of equal rank.

Under this new act all volunteers were to turn in their arms and equipment at the end of their enlistments and received a ten-dollar bounty. The men could no longer keep their weapons and equipment once they had been discharged.

The Act of 29 January 1813 created 20 new infantry regiments in the army modeled after the Voluntary Corps in which all members were enlisted for one year with all of the pay and benefits of the other regular soldiers except they were not entitled to land bounties.[11] Section 18 of this act repealed the acts of 6 February 1812 and 6 July 1812 preventing any new militia companies or regiments to be accepted into the Voluntary Corps. This act did not disband the corps. The corps would continue with its existing structure for another year. Any militiaman who wanted to serve with the army for only one year would have to enlist in one of these new regular army regiments. The repeal went into effect on 1 February 1813.

Two days later Congress passed another act giving non-commissioned officers and soldiers in both the militia and the volunteers, serving with the federal government, all the same monthly pay, rations and

---

[10] *Ibid*, Chapter CXXXVIII, page 785, 6 July 1812, "an act supplementary to the act authorizing the President of the United States to accept and organize certain Volunteer Military Corps."

[11] *Ibid*, Session II, Chapter XIV, pp. 794-798, 29 January 1813, "an act in addition to the act entitled, 'An act of raise an additional military force,' and for other purposes."

forage, and camp equipment as their counterpart in the army.[12] The expense of the militia and the volunteers was included in the army appropriations for 1813 under a bill, which was passed on 3 March 1813.[13] The Voluntary Corps would continue to be funded through the end of 1813.

The Voluntary Corps lasted for two years and 18 days when on 24 February 1814[14] Congress passed an act disbanding the corps. After 29 January 1813 the corps had became smaller in size through attrition and what was left of the remaining regiments had been consolidated into new volunteer regiments.

This new act incorporated the remaining companies and regiments of the Voluntary Corps into four new army regiments, the 45th through 48th Regiments of U.S. Infantry. These regiments could enlist men for five years or for the duration of the war. Many of the former volunteers now in the U.S. Army would re-enlist once their initial one-year service was completed.

All of the volunteers who were accepted into the four army regiments were given all of the pay and benefits associated with the rest of the army personnel. Officers would keep their rank and dates of commission once they were accepted into the army. The Act of 30 March 1814[15] permitted the officers of the former volunteer corps who were now on active duty with the army to be promoted in the line of the army, that is, these officers would be integrated into the army's officer promotion system.

In a turn of events, the Voluntary Corps was recreated on 27 January 1815 when Congress passed an act authorizing the President to accept into the service of the United States any corps of troops raised by the states for one year, not to exceed 40,000 men.[16] The President could also accept individuals into the corps. Each state was given a quota for the number of men that would be accepted into the corps.

**40,000-man Voluntary Corps**
**Quota of Men per State**

| Connecticut | 1,540 | New Jersey | 1,318 |
|---|---|---|---|
| Delaware | 440 | New York | 5,933 |
| Georgia | 1,318 | Ohio | 1,318 |
| Kentucky | 2,196 | Pennsylvania | 5,055 |
| Louisiana | 220 | Rhode Island | 450 |
| Massachusetts | 4,395 | South Carolina | 1,980 |
| Maryland | 1,980 | Tennessee | 1,318 |
| North Carolina | 2,858 | Virginia | 5,055 |
| New Hampshire | 395 | Vermont | 1,318 |

Once called to duty, these volunteers would receive the same pay and benefits (except land bounties) as the regular army counterparts. The President would commission the officers and they would also receive the same pay and benefits (except land bounties) as the regular army counterparts. Any non-commissioned officer or soldier who served for not less than two years was entitled to the 160-acre land bounty. The heirs of those men who were killed or who had die in the line of duty would also receive the

[12] *Ibid*, Chapter XVIII, page 797, 2 February 1813, "an act supplementary to an ant entitled, 'an act to provide for calling forth the militia to execute the laws, suppress insurrections, and repel invasions, and to repeal the act now in force for those purposes, and to increase the pay of volunteer and militia corps."

[13] *Ibid*, Chapter LVI, page 822, 3 March 1813, "an act making appropriations for the support of the military establishments and of the volunteer militia in the actual service of the United States, for the year one thousand eight hundred and thirteen."

[14] *Ibid*, Chapter XVI, page 98, 24 February 1814, "an act to authorize the President to receive into service certain volunteer corps."

[15] *Ibid*, Chapter XXXVII, pp. 113-116, 30 March 1814, "an act for the better organizing, paying, and supplying the army of the United States."

[16] *Ibid*, Session III, Chapter XXV, pp. 193-195, 27 January 1815, "an act to authorize the President of the United States to accept the services of state troops and of volunteers."

land bounty. The Voluntary Corps would be financed through the appropriations authorized for the militia units on federal duty for 1815.

This act lasted for only a month, for after the peace treaty was ratified ending the war, the act was repealed on 27 February 1815.[17] It is doubtful that any provisions of this act re-creating the Voluntary Corps were ever initiated. The Act of 3 March 1815 downsized the army to pre-1812 levels keeping eight infantry regiments on federal duty.[18] The Voluntary Corps and its offshoot of the 20 regular army infantry regiments, which were created in 1813 and the four regular army infantry regiments, which were created in 1814, were disbanded.

## The Missions

The U.S. Voluntary Corps was created to supplement the U.S. Army until each of the existing army regiments were at full strength. Initially, most of the volunteer companies were attached to individual army regiments.

Many of these companies would serve as light infantry or as rifle units within the various army infantry regiments that were created during 1812. Other independent companies would be organized as volunteer regiments and provide the same services to army brigades. The corps also provided artillery and light dragoon companies for the army.

Finally, volunteer regiments were created in New England in order to man the various forts and posts, which freed up regular army regiments so that they could be deployed to the front lines. The governor of Massachusetts (which also included the District of Maine) and the governor of Connecticut refused to call up the state's militia in support of the war. These states went so far as to discourage its citizens from joining the army and the Voluntary Corps. Most of New England was anti-war, which made it difficult for the army to enlist men for both the regular regiments and the Voluntary Corps.

Three volunteer companies, two from Pennsylvania and one from Virginia, were attached to the 19th Regiment of U.S. Infantry in Ohio where they provided both rifle and light infantry support for this regiment. Another volunteer company from Virginia was attached to the 20th Regiment of Infantry and it participated in the St. Lawrence campaigns of 1813.

Part of a composite volunteer regiment, made up of companies from Maryland, New York and Pennsylvania, along with a company from the 1st Regiment of U.S. Rifles, spearheaded the invasion of York, Upper Canada, securing a beachhead so that the regular army could land its troops without opposition.

A volunteer regiment formed in the District of Maine manned a number of forts in central Maine and they were also used to help stop the smuggling of goods between New England and the maritime colonies of Canada.

The U.S. Voluntary Corps' list of accomplishments is long and it has been greatly overlooked. The corps participated in all of the campaigns in the war and they saw action primarily in Louisiana, Mississippi, New York, and Ohio.

## The Problems with Researching

Identifying which militia units were traditional militia and which were under U.S. control as part of the Voluntary Corps can be challenging. Many of the traditional militia units used the term 'volunteers' in their designation while many of the Voluntary Corps units used their states names in their designation. Most Voluntary Corps units however used 'U.S. Volunteers' in their official designation.

---

[17] **Ibid**, Chapter LXIV, page 219, 27 February 1815, "an act to repeal certain acts therein mentioned."

[18] **Ibid**, Chapter LXXIX, pp. 224-225, 3 March 1815, "an act fixing the military peace establishment of the United States."

The best identifier is to have a muster or roster showing the dates of enlistment and the discharge dates. If the difference between the two dates is one year then that unit was a part of the Voluntary Corps. However, some Voluntary Corps units were released early from duty.

The records at the National Archives are not consolidated for both individuals who served in the war or for units who were called for a tour of duty.[19] The Voluntary Corps records can be found in both the U.S. Army records and the militia records for the War of 1812. Many of the ranger companies and sea fencibles companies formed by the regular army were also labeled at 'volunteers' even though these companies were not militia or volunteer units.

A second identifier is the officer's ranks. The U.S. Voluntary Corps was modeled after the U.S. Army's organization and therefore the officer's ranks within a company consisted of a captain, a first lieutenant, a second lieutenant, a third lieutenant and an ensign. The militia, as prescribed by the Militia Act of 1792 and by the various state and territorial militia laws, only had a captain, a lieutenant and an ensign as officers of a militia company.

The last identifier is race. The Militia Act of 1792 and most state and territorial laws permitted only white males from participating in the militia. The U.S. Voluntary Corps, on the other hand, not only had white males but Blacks, Indians, pirates and Canadians serving within its ranks. The U.S. Voluntary Corps was the first military organization in the United States to have Blacks and Indians commissioned as officers under the authority of the President.

## After the War

The traditional militia organization within the United States never evolved into the National Guard and the Reserve forces that are a part of our current military establishment. The militia was an involuntary military organization that had eventually died out by end of the American Civil War.

The War of 1812 was the last major war in which the traditional militia served. During the Mexican-American War a few militia companies were called up early in the fighting but it was soon realized that the short periods of time in which these militiamen wanted to serve would not be sufficient since most of the fighting was deep in Mexican territory. These militia companies were released early from duty.

The federal government once again established a voluntary corps in order to supplement the army during the Mexican conflict. Each state was given a quota of men needed for the war effort. Many of these regiments that were formed from the state volunteers gave superior service to our country.

Between the War of 1812 and the Civil War the traditional militia became a voluntary force while those men who did not want to serve were exempted and placed into the unorganized militia. The voluntary militia from each state and territory were renamed as the National Guard during the Civil War period.

---

[19] *Records of Volunteer Soldiers Who Served in the War of 1812*, Record Group 94,
http://www.archives.gov/publications/microfilm_catalogs/military/military_service_records_part04.html

# A War of 1812 roster has a unique tie to history

Researching can have many rewards. Whether you are researching a family, finding material for a book or an article, or just gaining additional knowledge on a certain subject, you may come across something that is very unique and worth sharing with others.

Recently I was helping a Canadian prepare an article which involved the War of 1812. She needed information on an Ohioan who had served during the War of 1812 and who had moved to Canada after the hostilities had ended. This ex-soldier was the subject of her article.

Out of courtesy, she sent me a copy of a document containing a list of American military officers who were held as prisoners of war in Canada during the War of 1812. She had seen this list in the Library and Archives Canada (the equivalent to our National Archives) and she had thought that I might be interesting in having a copy.

After first reading this list of names, I found that both of my eye brows had risen a couple of inches above my forehead. The list not only included Ohioans but it also included some very prominent Americans who would mold the U.S. Army after the War of 1812. It also included men who would have an impact on both American and Ohio politics between the War of 1812 and the American Civil War. If one or any of these men on this list had died in a British prisoner of war camp in Canada, the course of American history could have been altered.

The list of Americans included the names of twenty-two militia and regular army officers who have been captured after the fall of Fort Detroit and during the Battle of Queenston in 1812. The first year of the war went extremely bad for the Americans who had many more defeats than victories on the battlefield.

One of our victories was then Captain David Porter of the U.S. Frigate *Essex* captured the British transport *Samuel and Sarah* on 24 June 1812 on the Atlantic Ocean. This transport was bringing part of a battalion of the 1st Royal Scots Regiment of Foot to fight for the Canadians. This single act would set the stage for a British-American prisoner of war exchange on 17 September 1813.

The twenty-two officers were being exchanged for 159 British soldiers from the 1st Royal Scots. This may seem a little lop-sided, but the exchanged system was based upon a point value. A private was worth one point, a captain six points and a brigadier general twenty points. Each military rank was assigned a certain number of points.

The Americans on this list were one brigadier general, three colonels, four lieutenant colonels, one major, ten captains and three lieutenants for a total of 185 points. The British on their list had three lieutenants, nine sergeants, six corporals, two drummers and 139 privates for a total of 185 points. The men were exchanged on paper. Most of the officers had been paroled after their capture and they had been sent home to their regiments.

The four Ohioans were Duncan McArthur of Chillicothe, James Findlay of Cincinnati, Lewis Cass of Zanesville and Robert Lucas from the Portsmouth area.

Colonel Duncan McArthur was the commander of the 2nd Ohio Volunteer Regiment under Brigadier General William Hull and he was also a major general in the Ohio Militia. After his release, he was promoted to brigadier general in the U.S. Army and he would take command of the Army of the Northwest upon the resignation of Major General William Henry Harrison in 1814. McArthur would become the 11th governor of Ohio.

Colonel James Findlay was the commander of the 1st Ohio Volunteer Regiment under General Hull and he was also a brigadier general in the Ohio militia. He had been the mayor of Cincinnati before the War of 1812 and he would become a U.S. Congressman after the war.

Colonel Lewis Cass was the commander of the 3rd Ohio Volunteer Regiment under General Hull and also he was a brigadier general in the Ohio militia. After his release, he was promoted to brigadier general in the U.S. Army. He would resign from the army in late 1813 to become the 2nd governor of the Territory of Michigan. Later, he became an American ambassador to France, a U.S. Senator from Michigan, a Secretary of State and a Secretary of War. He ran for president of the United States in 1848.

Captain Robert Lucas was also a brigadier general in the Ohio militia. After the war, he would become a major general in the Ohio militia, the 12th governor of Ohio, and later, the 1st governor of the Territory of Iowa.

The other prominent men on the list of twenty-two were William Hull, Abraham Hull, James Miller, Winfield Scott, John Whistler, Josiah Snelling, Thomas Sidney Jesup, and Henry Brevoort.

Brigadier General William Hull was the commander of the Army of the Northwest and he surrendered his army to the British on 16 August 1812 at Fort Detroit without a fight. He is considered to be the "Benedict Arnold" of the War of 1812. Hull was the uncle to Commodore Isaac Hull, the noted commander of the U.S. Frigate *Constitution*.

Captain Abraham Fuller Hull was the son and aide-de-camp to General Hull. He would be assigned to the 9th Regiment of U.S. Infantry after his return to the United States. He would be killed during the Battle of Lundy's Lane on 25 July 1814. He has a marked grave on this battlefield in Ontario, Canada.

Lieutenant Colonel James Miller was the acting commander of the 4th Regiment of U.S. Infantry under General Hull. He had fought in the Battle of Tippecanoe in 1811 and he would become the commander of the 21st Regiment of U.S. Infantry. He would lead this regiment in the Battle of Lundy's Lane. After the battle he was promoted to brigadier general in the army. He would be elected as a U.S. Representative and later he became the first governor of the Territory of Arkansas.

Lieutenant Colonel Winfield Scott served in the U.S. Army for 47 years. He was the commanding general of the U.S. Army during his last 20 years in the army. He commanded the U.S. forces during the Mexican-American War and served as the military governor of Mexico City. He ran for the presidency in 1852.

Scott molded the Union strategy during the American Civil War when President Lincoln adopted his Anaconda Plan. This plan called for the U.S. Navy to blockade the Southern ports and the U.S. Army to advance down the Mississippi River to split the Confederacy in two. The plan worked!

Captain John Whistler's story is very interesting. A native of England, he enlisted in the British Army and served under Major General John Burgoyne during the American Revolution. He fought at the Battle of Saratoga and then immigrated to the U.S. after the war. He is the father of James Whistler, the famous American artist.

Captain Josiah Snelling would end his military career as a colonel is the U.S. Army. He is noted for building Fort Saint Anthony on the upper Mississippi River. This fort would protect the early settlers in the area that would become Minnesota.

First Lieutenant Thomas Sidney Jesup would become the "Father of the Modern Quartermaster Corps" for the U.S. Army. He served in the army from 1808 to 1860 rising to the rank of brigadier general. He had an Ohio connection! He was transferred to Ohio's 19th Regiment of U.S. Infantry after his release and he was in charge of building the small boats needed by General Harrison for his invasion of Upper Canada in late 1813. He was stationed in Cleveland, Ohio.

Captain Henry B. Brevoort of the 2nd Regiment of U.S. Infantry would command the marine detachment on the U.S. Brig *Niagara* during the Battle of Lake Erie on 10 September 1813.

This list of twenty-two American military officers has been known for two hundred years. The list is available to anyone through the British, the Canadian or American national archives. What makes this list interesting is to have been able to research each officer to see what that officer contributed in the War of 1812 and afterwards.

# The American Regimental System

The American regimental system for the U.S. Army that was used during the War of 1812 is a hybrid system that was developed from the British regimental system, the regimental system that was found on the continent of Europe (the Continental regimental system), and the need for the young nation to adapt to the style of Indian warfare found in the new world.

It is hard to compare the American system to the British system and to the continental system simply because all three systems were developed for a particular need. Up until the Revolutionary War the regimental system of the colonial militia was a direct adaptation of the British militia system in which each county supported a regiment, which fluctuated in the number of companies and the size of the companies during peace time and during war time. Larger counties could support more than one regiment.

An American army regiment tended to be smaller in troop size than its European counterpart because the command and control of a smaller size regiment was ideal for the vast woodlands of the American continent. In most cases a regiment of 500 men was large enough to combat any Indian insurgency. Also, the colonies did not have the large open spaces in which large armies, let along a European style regiment, could maneuver properly.

The American regiments were top heavy with officer ranks compared to the officer ranks in British or European regiments. During the War of 1812 a colonel commanded an American regiment, which had approximately the same number of men as a British or European battalion. These battalions were commanded by a lieutenant colonel. A major commanded the American battalions, which at most times were less than half the strength of a British or European battalion.

In order to understand the differences between the three systems, we must first do a comparison of the British System and the Continental System. Both systems used the regiment as the administrative unit and the battalion as the operational and tactical unit. Battalions were divided into companies. Companies were administrative units but when in combat or in formation companies used platoon tactics.

The number of companies in each of the systems varied with the branch of the service. The infantry tended to have between six and ten companies. If a regiment had six companies assigned through their authorization, then these companies would have had a larger troop strength than a regiment with ten companies. Cavalry regiments varied between six and eight companies while artillery regiments could have up to 20 companies assigned. There are many variations to the size of all types of regiments due to the unit's mission and recruiting efforts.

## The British regimental system

The British system was developed to support a small standing army, that is, during peacetime each regiment would only have one battalion with approximately ten companies assigned. During wartime, the regiments added more battalions when needed, which were recruited from the county from where the regiment originated. During the 1700's and through the War of 1812 most British regiments maintained only one battalion, thus, the terms 'regiment' and 'battalion' can be used interchangeably when referring to this system. Very few regiments ever raised a second battalion during this time period. The strength of a British regiment lay in its recruiting potential.

The noted exception is the 60th Regiment of Foot, which was raised during the French and Indian War with four battalions while during the War of 1812 this regiment had seven battalions. This regiment is the equivalent of today's French Foreign Legion and it was made up of mostly foreign volunteers, mainly Germans.

The British infantry regiment was made up of eight standard infantry companies, one light infantry company and one grenadier company. The British also had specialized regiments for light infantry and for rifles. Most infantry regiments were at half strength throughout most of the War of 1812.

Many regiments recruited their first battalions to full strength and they left a small core of officers and non-commissioned officers at home, which became the recruiting service to organize a second battalion when needed. Rarely did the second battalion of a regiment serve with its sister battalion. By the time the second battalion was organized the first battalion was already in combat. During wartime, three battalions from three different regiments would be combined to form a brigade. Thus, a British brigade was equivalent to a continental regiment, each having approximately the same number of troops.

## The continental regimental system

The continental system was developed to support a large standing army. European countries maintain from two to four battalions per regiment, and therefore, they were from two to four times larger than both the British and the American regiments. The British regiments could recruit up to the strength of the other European armies but the Americans keep the size of their regiment's constant and then expanding would organize new regiments. Unlike their British counterpart, a European regiment with all of its battalions operated for the most part as a unified command and fought together as a regiment.

The country with the greatest influence on the American regimental system was France. France had been an ally of the United States during the Revolutionary War and as such supplied the Continental Army with advisors, arms, uniforms, and supplies. Even thirty years after the war ended, France was still our major supplier of uniform parts, swords and other weapons. The American muskets were of a French design. All of our cadets at the Military Academy took French lessons. The army had been using the 1792 drill manual from the French army up until 1812 when a new version of the drill manual was issued showing the latest tactics of the French army during the Napoleonic Wars. We pronounce the rank of lieutenant as 'loo ten ant' as the French pronounce the rank instead of the British pronunciation of 'left ten ant.'

Up until 1806 Napoleon's infantry regiments had four battalions consisting of three war battalions and a depot battalion. The three war battalions operated together while the depot battalion was a recruiting battalion stationed back home in one of the French provinces. The war battalions was made up of seven standard infantry companies, a light infantry company, and a grenadier company, which was used as stock troops. A fully recruited regiment would have had approximately 3,600 men plus up to 400 men in the depot battalion.

In 1808 Napoleon reorganized the infantry regiments into two war battalions and a depot battalion. The third war battalion was disbanded and the men sent to the remaining two war battalions. The number of companies in a war battalion dropped from nine to six companies but the size of the company was increased. With this new organization the regiments became smaller with the maximum number of men in the war battalion set at 1,280 soldiers. A major commanded a depot battalion and it consisted of four infantry companies. Once the recruits were trained, they were sent to one of the two war battalions as replacements.

## The American regimental system

The American regimental system is actually two different organizations, one for the regular army and one for the militia. Surprisingly, the British-American militia system, which existed before the Revolutionary War is the basis of the American regular army and for the most part would be intact until the World War I era. With few exceptions, the army's infantry regiments would consist of ten companies until after the American Civil War.

The American militia regiments, however, were developed from the early militia's experiences during the Indian conflicts of the 1700's. The militia used a 'legion' approach to its regimental structure.

### The American militia regiments

The Militia Act of 1792 established a uniform militia organization within the United States and it was passed in two parts, which was approved by Congress on 2 May 1792 and on 8 May 1792. These acts set the standards on how the states and territories were to organize their militia and how these militia organizations were to be used in both peace and war.

The militia act used the legion approach in organizing the militia. A legion was a military unit made up of more than one branch of the army. These units would consist of infantry, cavalry and artillery companies unified under one command. The infantry companies could be standard infantry, light infantry or rifle infantry. The cavalry companies could be regular cavalry, light dragoons, or dragoons (mounted infantry). The artillery companies could be heavy artillery, used in fortifications, or light artillery which was used on the battlefield.

The federal government required each state and territory to divide its militia into divisions, brigades, regiments, battalions and companies. A division would be divided into two brigades with an attached artillery company and a cavalry company. Each brigade would consist of four infantry regiments while each infantry regiments would have five infantry companies and one rifle company attached.

Many of the states and territories modified their militia structure but they kept the basic idea of a legion. Most states and territories in the west could not afford to finance the required number of artillery and cavalry companies. Mounted infantry companies (dragoons) were used in the place of standard cavalry. These companies used horses for transportation but they normally dismounted during a battle and fought on foot. In the east, the states had a better financial base to support artillery and light dragoons companies. Due to the terrain of the United States, traditional cavalry companies were not formed.

During the War of 1812 the federal government did not activate existing militia brigades or regiments. Instead, new militia units were organized from existing units. If a militia brigade was needed for duty then a division was tasked to form a provision brigade made up of volunteers, and at times drafted men, from all of the brigades within a division. This approach ensured that no one county within a division was without local militia support.

The strength of the militia in the west was in the provisional brigades, which were normally made up of at least two infantry regiments and artillery, rife and mounted infantry companies as needed. In the east, infantry, artillery and mounted infantry regiments, battalions and companies were called up as needed to supplement the regular army.

## The American army regiments

The regimental structure of the U.S. Army during the War of 1812 consisted of ten standard infantry companies and a recruiting service. When a regiment was first authorized, all of the officers were commissioned and the senior non-commissioned officers (NCO) were enlisted, and they were all assigned to the regiment's recruiting service. The nation was divided into 48 recruiting districts and each regiment was assigned to recruit in one or more districts.

Once 100 men were recruited and sent to the regimental recruiting depot, the officers and NCOs were appointed from the recruiting service and a new company was formed. As the regiment grew, the recruiting service became smaller. Under the Act of 3 March 1813 a second major's position was authorized for an infantry regiment plus ten third lieutenants and ten additional sergeants. These new men were assigned to a regiment's recruiting service thus keeping this service alive and functioning.

By law, the infantry regiments were not battalions nor they were not divided into battalions, but battalions of two or more companies were formed "ad hoc" to perform certain missions. Normally, a major or a brevet major commanded these ad hoc battalions. Likewise, there were no official unit called a platoon in the American army, but companies were divided into two platoons when in formations, in marching, or in a line of battle. The captain commanded one platoon while the first lieutenant commanded the other platoon.

The army did have regiments of rifle, artillery and light dragoons but these organizations never operated as a single unit. Companies or ad hoc battalions were assigned to the various army brigades. Militia companies were used to perform the light infantry functions and these companies were attached to either a regular infantry regiment or a brigade.

Congress, under the Act of 11 January 1812, authorized ten new infantry regiments, which were organized similarly to France's regimental system. These regiments were organized with a headquarters' staff and 18 companies of 114 men each arranged in two battalions for a total of 2,016 men. The act of 26 June 1812 established the manpower strength of all infantry regiments at 1,070 men in ten companies of 109 men plus a headquarters' staff. The second battalion of each of the ten regiments was split off to become their own regiments on this same date.

The only American unit authorized during the War of 1812 to have battalions was the Corps of Artillery, which was created on 12 May 1814 when the 1st through 3rd Regiments of Artillery were combined into a single unit. This new artillery corps had a manpower strength of 5,916 men arranged into 12 battalions with four companies each.

A British battalion cannot be compared to an American regiment simply because a British battalion was a tactical organization while the American regiment was both an administrative and a tactical organization. The British battalion was commanded by a lieutenant colonel while a colonel commanded the American regiment. The American regiment system is probably closer in comparison to the French system than the British system but with a smaller manpower strength.

## Summary

There is no direct correlation between the regiments of the three systems. Each was designed for a specific purpose and each is unique. The British regimental system was designed for a small standing army and it was dynamic enough to increase in size to meet any crisis. The continental regimental system was designed for large standing armies and they were fully organized both in peace and in war. The American regimental system was designed for a small battlefield and its authorized size remained constant throughout a conflict.

The British brigade of three infantry regiments (battalions) was equal in strength to a French infantry regiment. An American regiment was equal in manpower to a British battalion but its command structure was stronger than that of the British's.

The tactical strength of all three regimental systems was in the division, which was usually made up of two or more infantry brigades, and companies or regiments of rifle, cavalry and artillery as needed. The American did not use the term 'division' for the units that were commanded by a major general. Instead, these units were called an 'army' as in Major General William H. Harrison's Army of the Northwest.

The charts below show the comparison of American, British and French military units and their manpower strengths between 1812 and 1815. Not all companies were at full strength so the manpower levels may vary. The number of men in each type of unit is an estimate.

| Division | | |
|---|---|---|
| **British** | **French** | **American** |
| Three infantry brigades - artillery, rifle and cavalry companies as needed 7,000 men | Two infantry brigades - artillery, rifle and cavalry companies as needed 12,000 men | Two infantry brigades - artillery, rifle and cavalry companies as needed 4,000 to 5,000 men |

| Brigade | | |
|---|---|---|
| **British** | **French** | **American** |
| Three battalions 3,000 men | Two regiments 5,120 men | Two regiments 2,000 men |

| Regiment | | |
|---|---|---|
| **British** | **French** | **American** |
| One or two battalions 1,000-2,000 men | Two war battalions 2,560 men | Ten companies 1,000 men |

| Battalion | | |
|---|---|---|
| **British** | **French** | **American (ad hoc)** |
| Ten companies 750-1000 men | Six companies 1,280 men | Two to five companies 150 to 400 men |

## Bibliography

*Battalion Organisation during the Second World War*, The British Regimental System,
http://www.bayonetstrength.150m.com/stuff/british_regimental_system.htm

*Histoire et Figurines, Premier Empire - Napoleon*, Armies – Angelterre/England, English Troops and Divisional Organization,
http://www.histofig.com/history/empire/

*Land Forces of Britain, the Empire and Commonwealth*, Regiments and Corps, Dictionary of Unit Nomenclature,
http://www.regiments.org/regiments/nomencla.htm

Muir, Roy, *Tactics and the Experience of Battle in the Age of Napoleon*, (New Haven, Connecticut and London, England: Yale University Press, 1998).

*Napoleon, His Army and Enemies*, Armies, Campaigns, Battles, Tactics, Commanders, French Infantry During the Napoleonic Wars, Organization. Regiments, Battalions and Companies,
http://web2.airmail.net/napoleon/infantry_Napoleon.html#frenchinfantryorganization

*Public Statues at Large of the United States of America*, volume I, (Boston, Massachusetts: Charles C. Little and James Brown, 1845).

Rogers, H.C.B., *Napoleon's Army*, (South Yorkshire, England: Pen & Sword Books Limited, 1974).

# Invalid Pensioners Living in Ohio during 1850

After every major war from the American Revolution through the Civil War, Congress published the list of soldiers who were on the invalid pension list. These men were either wounded in battle or were injured in the line of duty while serving in the U.S. Army, the volunteers or the militia. The Invalid Pension List of 1850 [20] was actually constructed in 1849 and published the following year.

There are 324 men on this list who were drawing their pensions from banks within Ohio. Some of these men were living in adjoining states. A few of these men probably served from other states and had moved to Ohio after their service ended in the military. Other men who served from Ohio moved to other states once their service ended and they would be listed on the pension list of the state in which they were living in 1849.

The list shows the name of the veteran and his rank plus the county in which he was living in during 1849. The next column has the date on which the veteran first drew his pension. This is an indication of the conflict in which he served. Dates between 1846 and 1848 would indicate Mexican War service while dates between 1812 and 1815 would represent War of 1812 service. The other dates may indicate service during the many Indian wars of this period. The yearly amount of the veteran's pension is shown in the next column while in the last column, "Mil. Est." represents veterans from the U.S. Army while dates represent militiamen.

"Mil. Est." means "Military Establishment," which refers to the regular army. The regular army had their own pension program which was separate from the volunteers and militiamen but having the same benefits. The volunteers and militiamen would qualify for their pensions by Acts of Congress which are dated in the last column.

Not shown on this list are the naval pensioners. These men have a separate published pension list. However, there is one U.S. Marine on this list. Also not included on this list are the heirs of the men who were killed or who had died in the line of duty. These heirs usually received the half pay of their love ones for five years. During the War of 1812 some of the heirs could elect to receive land bounties instead of half pay. Technically, the half pay recipients did not receive a pension.

The final group of veterans who are not on this list are the Revolutionary War veterans who were receiving service pensions. These pensions were issued to veterans for their service, not their wounds.

Once a veteran was totally recovered from his injury or wounds, he could lose his pension. The veterans would have been self-supporting in order for this to happen. Pensions during this era were designed to get a veteran back on his feet. Some men did received their pensions for life.

## Rank Designations

Pvt = private     Corp = corporal     Sgt = sergeant     Capt = captain     Maj = major
Matross and Artificer are enlisted ranks in an artillery regiment.

"Statement containing the names of all the invalid pensioners paid in the State of Ohio, the town or county in which they reside, the time when their pension commenced, the annual amount of pension received by each, distinguishing each according to grade, with reference to the several acts of Congress under which said pensions are allowed: prepared in conformity with the resolution of the House of Representatives of the 3d of March, 1849."

---

[20] *Invalid Pensioners*, Letter from the Secretary of the Interior, 31st Congress, 1st Session, House of Representatives, Ex. Doc. No. 74, 22 July 1850.

| Name | Rank | Residence | Commencement of pension | Annual amount received | Act under which pension was granted |
|------|------|-----------|------------------------|------------------------|-------------------------------------|
| Abbott, Elisha | Pvt | Hamilton Cty | 17 Apr 1846 | 96.00 | 24 Apr 1816 |
| Adam, William G. | Corp | Coshocton Cty | 13 Jan 1848 | 96.00 | Mil. Est. |
| Adams, Alanson | Pvt | Knox Cty | 31 Mar 1815 | 96.00 | Mil. Est. |
| Albers, John H. | Pvt | Hamilton Cty | 27 Nov 1849 | 96.00 | Mil. Est. |
| Alexander, Robert | Pvt | Harrison Cty | 19 Nov 1814 | 96.00 | 24 Apr 1816 |
| Ane, Adolphus | Pvt | Hamilton Cty | 4 Nov 1846 | 72.00 | Mil. Est. |
| Anthony, Patrick | Pvt | Hamilton Cty | 19 May 1848 | 96.00 | Mil. Est. |
| Armitage, William | Pvt | Muskingum Cty | 22 Sep 1837 | 96.00 | Mil. Est. |
| Armstrong, Andrew W. | 2 LT | Hamilton Cty | 14 Jun 1847 | 180.00 | 13 May 1846 |
| Ashury, Henry | Pvt | Highland Cty | 26 Dec 1812 | 48.00 | 24 Apr 1816 |
| Ausman, Abraham | Pvt | Clermont Cty | 1 Jan 1846 | 72.00 | 8 Aug 1846 |
| Austin, Horace | Pvt | Gallia Cty | 28 Apr 1838 | 96.00 | Mil. Est. |
| Auter, Thomas | Pvt | Hamilton Cty | 14 Dec 1827 | 48.00 | 6 Feb 1812 |
| Bacon, George | Marine | Lorain Cty | 17 May 1786 | 76.80 | 7 Jun 1785 |
| Baldwin, Thomas | Pvt | Hamilton Cty | 6 Jun 1815 | 96.00 | 30 Apr 1816 |
| Barney, Jonathan | Pvt | Montgomery Cty | 21 Oct 1848 | 96.00 | Mil. Est. |
| Barnum, Enoch | Pvt | Ashtabula Cty | 1 Jan 1817 | 96.00 | Mil. Est. |
| Batcheller, Benjamin | Matross | Cuyahoga Cty | 19 Mar 1816 | 32.00 | Mil. Est. |
| Bearnhardt, Amos | Pvt | Pennsylvania | 4 Jan 1848 | 96.00 | Mil. Est. |
| Beasley, Benjamin | Pvt | Brown Cty | 7 Apr 1849 | 96.00 | 25 Apr 1808 |
| Bennett, Thomas | Pvt | Hamilton Cty | 22 Feb 1843 | 96.00 | Mil. Est. |
| Bergen, Joseph | Pvt | Maryland | 19 Feb 1848 | 72.00 | Mil. Est. |
| Berrett, Henry | Pvt | Hamilton Cty | 5 Nov 1846 | 72.00 | Mil. Est. |
| Beverlin, Andrew | Pvt | Hamilton Cty | 7 Dec 1847 | 96.00 | Mil. Est. |
| Blake, Thomas | Pvt | New York | 10 Aug 1848 | 96.00 | Mil. Est. |
| Blinkensop, Thomas | Pvt | Hamilton Cty | 13 Jun 1847 | 96.00 | 13 May 1846 |
| Blodget, Samuel | Pvt | Montgomery Cty | 3 May 1847 | 96.00 | Mil. Est. |
| Bold, John | Sgt | Hamilton Cty | 14 Jan 1848 | 96.00 | Mil. Est. |
| Bold, John R. | Pvt | Miami Cty | 25 Oct 1823 | 96.00 | 24 Apr 1816 |
| Boone, Abner | Pvt | Harrison Cty, KY | 19 May 1848 | 72.00 | 24 Apr 1816 |
| Boree, Philip | Pvt | Ashtabula Cty | 7 Jun 1815 | 96.00 | Mil. Est. |
| Boughman, Daniel | Pvt | Fairfield Cty | 6 May 1848 | 96.00 | Mil. Est. |
| Bowen, Joseph | Corp | Summitt Cty | 16 Mar 1826 | 72.00 | Mil. Est. |
| Boyle, Andrew | Pvt | Hamilton Cty | 20 May 1833 | 96.00 | 6 Feb 1812 |
| Brafford, Thomas | Pvt | Brown Cty | 18 Oct 1843 | 96.00 | Mil. Est. |
| Brainard, Ashael | Drum Major | Lake Cty | 29 Nov 1842 | 240.00 | 4 Feb 1845 |
| Brannon, James | Corp | Portage Cty | 26 May 1831 | 96.00 | Mil. Est. |
| Brooks, John | Pvt | Erie Cty | 11 Feb 1825 | 57.00 | Mil. Est. |
| Brooks, Joseph | Corp | Monroe Cty | 4 Mar 1828 | 96.00 | Mil. Est. |
| Brown, John | Pvt | Hamilton Cty | 13 May 1847 | 96.00 | Mil. Est. |
| Bryant, Joseph | 2 LT | Lake Cty | 4 Sep 1814 | 180.00 | Mil. Est. |
| Buck, Henry | Corp | Summitt Cty | 9 Dec 1847 | 96.00 | Mil. Est. |
| Buckingham, Jared | Pvt | Cuyahoga Cty | 4 Mar 1832 | 64.00 | 14 May 1836 |
| Bundy, William | Pvt | Sandusky Cty | 26 May 1815 | 96.00 | Mil. Est. |
| Burt, John | Sgt | Cuyahoga Cty | 5 Dec 1818 | 32.00 | Mil. Est. |

| Name | Rank | Residence | Commencement of pension | Annual amount received | Act under which pension was granted |
|------|------|-----------|------------------------|----------------------|------------------------------------|
| Bush, Henry C. | Sgt | Erie Cty | 5 Jan 1848 | 64.00 | Mil. Est. |
| Bushnell, Andrew | 2 LT | Trumbull Cty | 16 Jun 1815 | 45.00 | Mil. Est. |
| Bussey, Herman | Pvt | Hamilton Cty | 29 Dec 1847 | 48.00 | Mil. Est. |
| Butler, John | Pvt | Hamilton Cty | 1 Jun 1847 | 48.00 | Mil. Est. |
| Byrnes, Timothy | Pvt | Hamilton Cty | 31 Oct 1847 | 96.00 | Mil. Est. |
| Cain, Robert | Pvt | Wayne Cty | 19 Dec 1835 | 96.00 | Mil. Est. |
| Caldwell, John W. | Pvt | Miami Cty | 17 Nov 1847 | 96.00 | Mil. Est. |
| Camm, James M. | Sgt | Cuyahoga Cty | 28 Oct 1847 | 48.00 | Mil. Est. |
| Campbell, James | Musician | Hamilton Cty | 14 Jan 1848 | 96.00 | Mil. Est. |
| Campbell, William P. | Pvt | Hamilton Cty | 3 Sep 1847 | 48.00 | Mil. Est. |
| Campbell, William P. | Pvt | Belmont Cty | 2 Dec 1847 | 64.00 | Mil. Est. |
| Cannon, William S. | Pvt | Crawford Cty | 31 Aug 1849 | 96.00 | 24 Apr 1816 |
| Carena, Charles | Pvt | Huron Cty | 2 Nov 1843 | 48.00 | Mil. Est. |
| Carle, Peter | Sgt | Hamilton Cty | 25 Apr 1846 | 96.00 | 24 Apr 1816 |
| Carlisle, John | Pvt | Unknown | 7 Nov 1849 | 48.00 | 13 May 1846 |
| Carpenter, Dorman | Pvt | Delaware Cty | 31 Oct 1847 | 64.00 | Mil. Est. |
| Caskey, John M. | Pvt | Hamilton Cty | 14 Jan 1848 | 48.00 | Mil. Est. |
| Catharell, Joseph | Pvt | Hamilton Cty | 19 Nov 1814 | 48.00 | 24 Apr 1816 |
| Cavan, Timothy | Pvt | Monroe Cty | 23 Jan 1844 | 96.00 | Mil. Est. |
| Clachan, Alexander | Pvt | Hamilton Cty | 9 Apr 1847 | 96.00 | Mil. Est. |
| Clark, Adam G. | Pvt | Belmont Cty | 15 Jan 1848 | 64.00 | Mil. Est. |
| Clark, John | Pvt | Hamilton Cty | 28 Nov 1847 | 96.00 | Mil. Est. |
| Clark, Samuel | Pvt | Medina Cty | 31 Aug 1841 | 72.00 | Mil. Est. |
| Cobb, Michael | Pvt | Hamilton Cty | 28 Nov 1847 | 96.00 | Mil. Est. |
| Coffee, George | Pvt | Columbiana Cty | 11 Feb 1848 | 96.00 | Mil. Est. |
| Cole, Samuel D. | Pvt | Mahoning Cty | 8 Feb 1847 | 96.00 | Mil. Est. |
| Coleman, William | Pvt | Montgomery Cty | 25 May 1847 | 64.00 | Mil. Est. |
| Copeland, Weeks | Corp | Delaware Cty | 18 Feb 1834 | 72.00 | 29 Jan 1813 |
| Copps, Josiah | Pvt | Burlington, KY | 21 Jan 1821 | 72.00 | Mil. Est. |
| Cox, William | Pvt | Fayette Cty | 21 Oct 1814 | 96.00 | Mil. Est. |
| Cral, Jeremiah | Pvt | Clark Cty | 13 May 1847 | 96.00 | Mil. Est. |
| Crany, Theodore | Pvt | Hamilton Cty | 19 Feb 1848 | 64.00 | Mil. Est. |
| Cremens, Moses | Pvt | Gallia Cty | 1 Jan 1832 | 72.00 | 2 Mar 1833 |
| Cressey, Moses | Pvt | Williams Cty | 17 Apr 1816 | 96.00 | Mil. Est. |
| Crubbs, Philip | Pvt | Jefferson Cty | 25 Feb 1832 | 32.00 | Mil. Est. |
| Culins, George | Pvt | Muskingum Cty | 3 Apr 1814 | 96.00 | Mil. Est. |
| Cunningham, Joseph | Pvt | Licking Cty | 3 Aug 1822 | 96.00 | Mil. Est. |
| Curry, Robert | Pvt | Hamilton Cty | 1 Jan 1828 | 96.00 | 20 May 1830 |
| Dains, Andrew | Pvt | Athens Cty | 14 Jan 1845 | 48.00 | Mil. Est. |
| Daniels, Benjamin | Maj | Pike Cty | 22 Feb 1814 | 300.00 | 18 Apr 1818 |
| Dawson, William | Pvt | Belmont Cty | 11 Jul 1819 | 96.00 | Mil. Est. |
| Denio, Frederick | Pvt | Ashtabula Cty | 17 Oct 1841 | 96.00 | Mil. Est. |
| Derr, Mathias | Pvt | Hamilton Cty | 6 Nov 1847 | 96.00 | 13 May 1846 |
| Devault, John | Pvt | Portage Cty | 21 Dec 1820 | 48.00 | Mil. Est. |
| Devon, Enos | Pvt | Perry Cty | 15 Jan 1830 | 72.00 | Mil. Est. |
| Dickson, Robert | Pvt | Holmes Cty | 20 Sep 1846 | 96.00 | 13 May 1846 |

| Name | Rank | Residence | Commencement of pension | Annual amount received | Act under which pension was granted |
|------|------|-----------|------------------------|------------------------|--------------------------------------|
| Donoghy, William | Pvt | Cuyahoga Cty | 13 Jan 1848 | 32.00 | Mil. Est. |
| Downs, Thomas | Pvt | Hamilton Cty | 10 Apr 1846 | 72.00 | Mil. Est. |
| Dukes, David | Pvt | Hamilton Cty | 30 Jul 1838 | 96.00 | Mil. Est. |
| Dunham, William | Corp | Erie Cty, NY | 16 Feb 1849 | 72.00 | 24 Apr 1816 |
| Dunn, William | Pvt | Greene Cty | 10 Feb 1848 | 72.00 | Mil. Est. |
| Dye, Daniel | Pvt | Hardin Cty | 11 Jun 1815 | 96.00 | 24 Apr 1816 |
| Eaton, Jeremiah A. | Pvt | Montgomery Cty | 15 Sep 1849 | 96.00 | Mil. Est. |
| Eaton, Origin | Pvt | Cuyahoga Cty | 28 May 1815 | 300.00 | 14 May 1838 |
| Eberling, William | Pvt | Hamilton Cty | 29 Feb 1848 | 72.00 | 13 May 1846 |
| Eckstein, Frederick | Pvt | Tuscarawas Cty | 2 Mar 1848 | 48.00 | Mil. Est. |
| Eldred, Moses | Pvt | Cuyahoga Cty | 24 Feb 1813 | 96.00 | 24 Apr 1816 |
| Elliott, Christopher | Pvt | Hamilton Cty | 10 Nov 1847 | 96.00 | Mil. Est. |
| Elliott, Francis | Pvt | Franklin Cty | 14 Apr 1844 | 96.00 | Mil. Est. |
| Enders, John | Pvt | Hamilton Cty | 28 Nov 1847 | 96.00 | Mil. Est. |
| Evans, George | Pvt | Hamilton Cty | 15 Jan 1848 | 96.00 | 13 May 1846 |
| Farrell or Fannell, James | Pvt | Hamilton Cty | 30 May 1846 | 96.00 | Mil. Est. |
| Fireobend, Soloman | Pvt | Richland Cty | 22 Jan 1822 | 96.00 | Mil. Est. |
| Fitzallen, James | Pvt | Hamilton Cty | 12 Nov 1848 | 96.00 | Mil. Est. |
| Fitzgerald, John | Sgt | Hamilton Cty | 25 Jul 1848 | 32.00 | Mil. Est. |
| Fleming, Michael | Pvt | Hamilton Cty | 22 May 1847 | 96.00 | Mil. Est. |
| Fletcher, John F. | Pvt | Hamilton Cty | 27 Nov 1846 | 48.00 | 13 May 1846 |
| Fling, Richard | Pvt | St. Louis, MO | 2 May 1815 | 64.00 | Mil. Est. |
| Flowers, Benjamin | Pvt | Brown Cty | 19 Nov 1814 | 48.00 | 24 Apr 1816 |
| Foot, James | Pvt | Cuyahoga Cty | 16 Oct 1840 | 48.00 | Mil. Est. |
| Foster, Lathrop | Pvt | Lake Cty | 1 Jan 1844 | 48.00 | 17 Jun 1844 |
| Fox, Frederick | Pvt | Hamilton Cty | 27 Sep 1849 | 72.00 | 13 May 1846 |
| France, John | Pvt | Hamilton Cty | 1 Jun 1847 | 64.00 | Mil. Est. |
| Francis, Joseph | Pvt | Sandusky Cty | 4 Apr 1848 | 96.00 | Mil. Est. |
| Frank, Jacob | Pvt | Trumbull Cty | 1 Dec 1812 | 32.00 | 24 Apr 1816 |
| Frick, Frank | Corp | Unknown | 9 Aug 1837 | 48.00 | Mil. Est. |
| Fugate, Thomas | Pvt | Clinton Cty | 31 May 1814 | 96.00 | 30 Apr 1816 |
| Fuller, Ambrose | Pvt | Rome, IL | 29 Jun 1815 | 32.00 | Mil. Est. |
| Gardner, Julius | Pvt | Cuyahoga Cty | 20 Nov 1846 | 96.00 | 24 Apr 1816 |
| Garrison, Joseph | Pvt | Hamilton Cty | 14 Jan 1848 | 48.00 | Mil. Est. |
| Gates, William | Pvt | Cuyahoga Cty | 2 May 1820 | 96.00 | Mil. Est. |
| Gilhurly, Joseph | Pvt | Mercer Cty | 20 Feb 1848 | 96.00 | Mil. Est. |
| Gipphard, William | Pvt | Hamilton Cty | 3 Feb 1848 | 48.00 | Mil. Est. |
| Goad, William R. | Pvt | Hamilton Cty | 3 Oct 1849 | 96.00 | Mil. Est. |
| Gorman, William | Pvt | Muskingum Cty | 25 May 1825 | 96.00 | 25 Apr 1808 |
| Gortner, Jesse | Pvt | Clark Cty | 6 Jul 1848 | 96.00 | Mil. Est. |
| Graham, John | Pvt | Hamilton Cty | 11 July 1848 | 48.00 | 13 May 1846 |
| Green, Elisha B. | Sgt | Lawrence, Cty | 8 Feb 1815 | 96.00 | Mil. Est. |
| Hall, Samuel | Pvt | Warren Cty | 11 Oct 1814 | 96.00 | Mil. Est. |
| Hamilton, Samuel C. | Corp | Franklin Cty | 6 Jul 1839 | 96.00 | Mil. Est. |
| Harper, William | Capt | Centre Cty | 16 Apr 1813 | 240.00 | 24 Apr 1816 |
| Havens, Joel | Pvt | Hamilton Cty | 4 Nov 1846 | 48.00 | Mil. Est. |

| Name | Rank | Residence | Commencement of pension | Annual amount received | Act under which pension was granted |
|---|---|---|---|---|---|
| Henn, Charles | 2 Lt | Hamilton Cty | 20 Dec 1848 | 90.00 | 13 May 1846 |
| Higgins, William | Pvt | Huron Cty | 31 Oct 1847 | 64.00 | Mil. Est. |
| Hill, John | Pvt | Hamilton Cty | 11 Jan 1847 | 48.00 | Mil. Est. |
| Hoefer, Henry (alias Andrew) | Pvt | Hamilton Cty | 6 Oct 1848 | 96.00 | Mil. Est. |
| Hogan, Daniel | Pvt | Hamilton Cty | 10 Nov 1847 | 48.00 | Mil. Est. |
| Hollister, Jesse W. | Sgt | Clermont Cty | 20 May 1846 | 96.00 | Mil. Est. |
| Holmes, James | Pvt | Hamilton Cty | 20 Aug 1846 | 96.00 | Mil. Est. |
| Honser, Henry | Pvt | Montgomery Cty | 1 Jan 1828 | 96.00 | 20 May 1830 |
| Hughes, William | Pvt | Hamilton Cty | 10 Aug 1848 | 96.00 | Mil. Est. |
| Isbell, Ransom | Sgt | Medina Cty | 9 Nov 1814 | 72.00 | 24 Apr 1816 |
| Jack, James | Pvt | Putnam Cty | 6 Apr 1832 | 72.00 | 24 Apr 1816 |
| Jackson, John | Pvt | Hamilton Cty | 29 Aug 1848 | 24.00 | Mil. Est. |
| Jackson, William | Pvt | Carroll Cty | 22 Nov 1834 | 96.00 | 24 Apr 1816 |
| Jennison, John S. | Corp | Hamilton Cty | 11 Jun 1847 | 96.00 | 13 May 1846 |
| Johnson, Abraham | Pvt | Cuyahoga Cty | 11 Feb 1816 | 96.00 | 30 Apr 1816 |
| Johnson, Arnold | Pvt | Hamilton Cty | 26 Oct 1849 | 96.00 | 13 May 1846 |
| Johnston, Harvey | Pvt | Cuyahoga Cty | 11 Apr 1815 | 96.00 | Mil. Est. |
| Jones, John B. | Corp | Hamilton Cty | 7 Feb 1850 | 96.00 | 13 May 1846 |
| Joyce, Michael | Pvt | Hamilton Cty | 16 Nov 1847 | 96.00 | Mil. Est. |
| Keller, John | Sgt | Scioto Cty | 23 Jan 1847 | 96.00 | Mil. Est. |
| Kelscher, Charles | Pvt | Defiance Cty | 4 Nov 1849 | 32.00 | 13 May 1846 |
| Kenney, Patrick | Pvt | Hamilton Cty | 16 Feb 1848 | 48.00 | Mil. Est. |
| Keyser, Samuel | Pvt | Hamilton Cty | 20 Apr 1849 | 48.00 | Mil. Est. |
| Killum, Josiah A. | Pvt | Clermont Cty | 11 Nov 1846 | 96.00 | 13 May 1846 |
| Kingsbury, William | Sgt | Stark Cty | 21 Aug 1815 | 96.00 | Mil. Est. |
| Knecht, Louis | Pvt | Wayne Cty, MI | 26 Nov 1847 | 64.00 | Mil. Est. |
| Knott, Michael | Pvt | Hamilton Cty | 28 Nov 1847 | 72.00 | Mil. Est. |
| Kuchner, William | Pvt | Franklin Cty | 25 Feb 1848 | 72.00 | 13 May 1846 |
| Laing, Lewis | Pvt | Butler Cty | 1 Jan 1846 | 96.00 | 8 Aug 1846 |
| Lane, Levi | Pvt | Trumbull Cty | 21 Jun 1815 | 96.00 | 24 Apr 1816 |
| Lane, Samuel | Pvt | Madison Cty | 29 Sep 1843 | 64.00 | 6 Feb 1812 |
| Lasear, John G. | Pvt | Muskingum Cty | 17 Feb 1848 | 72.00 | Mil. Est. |
| Leach, William | Pvt | Hamilton Cty | 14 Jan 1848 | 96.00 | Mil. Est. |
| Lee, Alfred | Pvt | Ashtabula Cty | 11 Nov 1814 | 72.00 | 24 Apr 1816 |
| Lilly, Anson | Pvt | Medina Cty | 1 Jul 1815 | 96.00 | 24 Apr 1816 |
| Lincoln, Mordecai | Pvt | Richland Cty | 18 Jun 1819 | 72.00 | Mil. Est. |
| Linden, Christopher | Pvt | Lorain Cty | 20 Feb 1848 | 64.00 | Mil. Est. |
| Lingrell, Thomas | Pvt | Ross Cty | 20 Jan 1841 | 32.00 | Mil. Est. |
| Linn, Erwin | Pvt | Fairfield Cty | 22 Feb 1848 | 48.00 | Mil. Est. |
| Love, Charles | Pvt | Hamilton Cty | 11 Aug 1815 | 84.00 | Mil. Est. |
| Lyon, Abraham | Pvt | Clark Cty | 27 Mar 1841 | 48.00 | 29 Jan 1813 |
| Lyon, Harvey | Sgt | Cuyahoga Cty | 4 Apr 1848 | 96.00 | Mil. Est. |
| Madden, John | Pvt | Hamilton Cty | 14 Jan 1848 | 96.00 | Mil. Est. |
| Maddox, Ebenezer | Sgt | Cuyahoga Cty | 2 Nov 1825 | 95.00 | Mil. Est. |
| Madegan, Michael | Pvt | Hamilton Cty | 31 Oct 1847 | 96.00 | Mil. Est. |

| Name | Rank | Residence | Commencement of pension | Annual amount received | Act under which pension was granted |
|------|------|-----------|------------------------|------------------------|--------------------------------------|
| Maitland, John | Pvt | Clark Cty | 22 Feb 1848 | 96.00 | Mil. Est. |
| Mann, Samuel | Pvt | Richland Cty | 20 Jan 1826 | 48.00 | 24 Apr 1816 |
| Mathers, Matthew A. | Pvt | Montgomery Cty | 26 Dec 1847 | 96.00 | Mil. Est. |
| Maxwell, William | Pvt | Greene Cty | 8 Oct 1816 | 96.00 | 3 Mar 1817 |
| McCloskey, Charles | Pvt | Coshocton Cty | 16 Feb 1848 | 96.00 | Mil. Est. |
| McConville, John | Pvt | Hamilton Cty | 30 Oct 1847 | 64.00 | Mil. Est. |
| McCourt, Bernard | Sgt | Hamilton Cty | 2 Jan 1850 | 96.00 | Mil. Est. |
| McDonald, John A. J. | Pvt | Hamilton Cty | 20 Feb 1849 | 48.00 | 13 May 1846 |
| McGartlin, Joseph | Pvt | Hamilton Cty | 19 Feb 1848 | 96.00 | Mil. Est. |
| McGuire, John | Pvt | Sandusky Cty | 6 May 1848 | 96.00 | Mil. Est. |
| McIntire, Alpheus | Sgt | Sandusky Cty | 14 Apr 1826 | 96.00 | 24 Apr 1816 |
| McIntire, Joseph C. | Pvt | Miami Cty | 6 Feb 1848 | 72.00 | Mil. Est. |
| McLaughlin, Robert | Pvt | Fairfield Cty | 12 Dec 1849 | 96.00 | 24 Apr 1816 |
| McNabb, John | Pvt | Fairfield Cty | 31 Jul 1838 | 96.00 | Mil. Est. |
| McNulty, James | Pvt | Hamilton Cty | 2 Aug 1848 | 96.00 | Mil. Est. |
| Meeker, Thomas J. | Pvt | Clermont Cty | 9 Jan 1850 | 96.00 | 13 May 1846 |
| Meins, John | Artificer | Hamilton Cty | 12 Jan 1848 | 96.00 | Mil. Est. |
| Metz, Charles | Pvt | Hamilton Cty | 5 Feb 1848 | 48.00 | Mil. Est. |
| Meyer, Henry | Pvt | Unknown | 27 Nov 1846 | 96.00 | 13 May 1846 |
| Mickle, Adolph | Pvt | Hamilton Cty | 24 Dec 1847 | 96.00 | Mil. Est. |
| Milburne, Andrew | Pvt | Huron Cty | 1 Mar 1831 | 96.00 | Mil. Est. |
| Miles, Jesse A. | Pvt | Hamilton Cty | 3 Feb 1843 | 96.00 | Mil. Est. |
| Miller, Charles | Pvt | Hamilton Cty | 15 Apr 1842 | 64.00 | Mil. Est. |
| Miller, George | Pvt | Hamilton Cty | 15 Oct 1849 | 96.00 | 13 May 1846 |
| Miller, Jacob | Pvt | Fairfield Cty | 22 Feb 1848 | 48.00 | Mil. Est. |
| Mills, Nathaniel | Pvt | Wayne Cty | 23 Feb 1843 | 48.00 | 24 Apr 1816 |
| Minshall, John | Pvt | Hamilton Cty | 2 Nov 1846 | 96.00 | 13 May 1846 |
| Mitchell, John B. | Pvt | Hamilton Cty | 6 Nov 1846 | 72.00 | 13 May 1846 |
| Momeny, George | Pvt | Sandusky Cty | 3 Apr 1848 | 96.00 | Mil. Est. |
| Montgomery or Springer, Shadrack | Pvt | Mercer Cty | 21 Oct 1848 | 96.00 | 24 Apr 1816 |
| Morrison, George W. | Pvt | Morgan Cty | 1 Jan 1832 | 96.00 | 20 May 1830 |
| Morse, Amos | Pvt | Geauga Cty | 1 Mar 1825 | 96.00 | Mil. Est. |
| Morton, Thomas | Pvt | Hamilton Cty | 10 Jan 1848 | 48.00 | Mil. Est. |
| Murphy, William | Pvt | Ashtabula Cty | 8 Aug 1845 | 96.00 | 24 Apr 1816 |
| Myers, Samuel | Pvt | Hamilton Cty | 15 Oct 1848 | 48.00 | 13 May 1846 |
| Nash, John | Pvt | Hamilton Cty | 31 Oct 1834 | 96.00 | Mil. Est. |
| Nichols, David | Corp | Pickaway Cty | 27 Aug 1824 | 96.00 | Mil. Est. |
| Noonan, Michael | Pvt | Hamilton Cty | 16 May 1847 | 96.00 | Mil. Est. |
| Nye, John | Pvt | Pickaway Cty | 1 Jun 1815 | 96.00 | Mil. Est. |
| Owens, Thomas | Sgt | Hamilton Cty | 17 Feb 1848 | 96.00 | Mil. Est. |
| Page, George A. | Pvt | Cuyahoga Cty | 10 Aug 1848 | 64.00 | Mil. Est. |
| Parks, George W. | Pvt | Licking Cty | 21 Mar 1815 | 72.00 | Mil. Est. |
| Patterson, Alexander | 2 Lt | Jefferson Cty | 16 Jun 1815 | 135.00 | Mil. Est. |
| Pierce, John | Pvt | Wayne Cty | 24 Apr 1816 | 72.00 | Mil. Est. |
| Piott, Lloyd | Pvt | Coshocton Cty | 23 May 1815 | 96.00 | Mil. Est. |

| Name | Rank | Residence | Commencement of pension | Annual amount received | Act under which pension was granted |
|---|---|---|---|---|---|
| Pitzer, Anthony | Maj | Licking Cty | 25 Jun 1836 | 225.00 | 24 Apr 1816 |
| Pompilly, Bernard | Pvt | Hamilton Cty | 1 Jan 1836 | 96.00 | 28 Jun 1836 |
| Porter, Bazlee | Pvt | Trumbull Cty | 1 Jan 1833 | 96.00 | 30 Jun 1834 |
| Rainey, Elijah | Pvt | Licking Cty | 16 Feb 1848 | 72.00 | Mil. Est. |
| Randall, John B. | Pvt | Unknown | 4 Dec 1813 | 96.00 | 8 Apr 1812 |
| Reed, Benjamin G. | Pvt | Medina Cty | 30 Jun 1835 | 72.00 | Mil. Est. |
| Reed, Henry H. | Pvt | Hamilton Cty | 11 May 1848 | 48.00 | Mil. Est. |
| Reed, James | Pvt | Hamilton Cty | 28 Apr 1846 | 48.00 | Mil. Est. |
| Regan, Thomas | Pvt | Hamilton Cty | 16 May 1847 | 96.00 | Mil. Est. |
| Reir, George | Pvt | Unknown | 25 Aug 1848 | 96.00 | Mil. Est. |
| Repp, John | Pvt | Stark Cty | 14 Aug 1812 | 48.00 | 16 May 1802 |
| Retman, Frantz | Pvt | Hamilton Cty | 27 Oct 1847 | 96.00 | Mil. Est. |
| Rice, Patrick H. | Pvt | Hamilton Cty | 1 Jun 1849 | 96.00 | Mil. Est. |
| Ritter, John D. | Pvt | Hamilton Cty | 30 Apr 1848 | 96.00 | Mil. Est. |
| Roberts, John A. | Sgt | Knox Cty | 31 Jul 1848 | 48.00 | Mil. Est. |
| Roper, Holly | Sgt | Clermont Cty | 7 Dec 1846 | 96.00 | 24 Apr 1816 |
| Ross, Joseph M. | Sgt | Champaign Cty | 31 May 1815 | 96.00 | Mil. Est. |
| Roth, Lewis | Recruit | Ross Cty | 9 May 1849 | 24.00 | Mil. Est. |
| Ruckstool, John | Pvt | Scioto Cty | 8 Jun 1847 | 96.00 | 13 May 1846 |
| Ruth, Nicholas | Corp | Hamilton Cty | 30 Jun 1843 | 48.00 | Mil. Est. |
| Ryneason, Minny | Pvt | Hamilton Cty | 6 Jan 1815 | 96.00 | 3 May 1815 |
| Salisbury, Andrew | Pvt | Greene Cty | 16 May 1816 | 96.00 | Mil. Est. |
| Sanders, Isaac D. | Sgt | Boone Cty | 28 Nov 1833 | 96.00 | Mil. Est. |
| Schinincke, Ludwig | Pvt | Hamilton Cty | 11 Apr 1848 | 96.00 | Mil. Est. |
| Schlesinger, Julius | Pvt | Hamilton Cty | 18 Nov 1847 | 48.00 | Mil. Est. |
| Schoupp, Frederick | Pvt | Montgomery Cty | 28 Jun 1848 | 96.00 | Mil. Est. |
| Schrofe, Emanuel | Pvt | Brown Cty | 23 Sep 1843 | 48.00 | 17 Jun 1844 |
| Schweigleman, Joseph | Pvt | Hamilton Cty | 22 May 1847 | 96.00 | Mil. Est. |
| Scribner, Salmasius B. | Pvt | Franklin Cty | 22 Aug 1848 | 48.00 | 13 May 1846 |
| Seals, Hezekiah | Pvt | Butler Cty | 20 Nov 1824 | 64.00 | Mil. Est. |
| Servess, William G. | 2 LT | Clark Cty | 28 Mar 1815 | 180.00 | 2 Jan 1812 |
| Seybold, John G. | Recruit | Hamilton Cty | 9 Apr 1848 | 72.00 | Mil. Est. |
| Shaddock, Martin | Pvt | Hamilton Cty | 14 Jan 1848 | 72.00 | Mil. Est. |
| Shane, Adam F. | Pvt | Hamilton Cty | 4 Dec 1846 | 96.00 | 13 May 1846 |
| Sharis, Jacob | Pvt | Hamilton Cty | 28 Oct 1847 | 96.00 | Mil. Est. |
| Shields, Hector | Sgt | Portage Cty | 6 Jul 1830 | 96.00 | Mil. Est. |
| Shiell, Robert O. | Corp | Hamilton Cty | 27 Oct 1847 | 96.00 | Mil. Est. |
| Shrier, Abraham | Corp | Unknown | 3 Aug 1844 | 64.00 | Mil. Est. |
| Sibberall, William | Pvt | Sandusky Cty | 7 Jun 1815 | 96.00 | Mil. Est. |
| Six, Conrad | Pvt | Morgan Cty | 6 Mar 1815 | 32.00 | Mil. Est. |
| Slocum, Thomas | Pvt | Knox Cty | 21 Oct 1836 | 84.00 | Mil. Est. |
| Smith, David | Pvt | Clark Cty | 12 Oct 1848 | 96.00 | Mil. Est. |
| Smith, John | Sgt | Washington Cty | 16 Dec 1848 | 96.00 | Mil. Est. |
| Smith, Richard | Pvt | Coshocton Cty | 30 May 1824 | 96.00 | Mil. Est. |
| Smith, William Jr | Pvt | Champaign Cty | 21 Dec 1838 | 48.00 | Mil. Est. |
| Snyder, William | Pvt | Columbiana Cty | 11 Jun 1815 | 96.00 | Mil. Est. |

| Name | Rank | Residence | Commencement of pension | Annual amount received | Act under which pension was granted |
|---|---|---|---|---|---|
| Soule, James | Pvt | Trumbull Cty | 4 Aug 1848 | 72.00 | Mil. Est. |
| Spears, William G. | Pvt | Franklin Cty | 10 Nov 1847 | 48.00 | Mil. Est. |
| Spohn, John | Pvt | Montgomery Cty | 4 Nov 1846 | 72.00 | Mil. Est. |
| Spoonover, James | Pvt | Scioto Cty | 25 Aug 1820 | 96.00 | 24 Apr 1816 |
| Stagg, Daniel | Pvt | Lebanon Cty | 26 Feb 1816 | 96.00 | 30 Apr 1816 |
| Stevens, Elhaman | Pvt | Pennsylvania | 7 Apr 1848 | 48.00 | 10 Aug 1848 |
| Steward, Marvin | Pvt | Greene Cty | 4 Apr 1848 | 48.00 | Mil. Est. |
| Stewart, Aaron | Pvt | Clark Cty | 2 Oct 1814 | 96.00 | Mil. Est. |
| Still, Samuel | Sgt | Brown Cty | 8 Nov 1814 | 96.00 | Mil. Est. |
| Stoddard, David | Pvt | Scioto Cty | 23 Apr 1815 | 96.00 | Mil. Est. |
| Stollings, Martin | Pvt | Jackson Cty | 17 Feb 1848 | 72.00 | Mil. Est. |
| Storm, John | Corp | Hamilton Cty | 20 Feb 1848 | 96.00 | Mil. Est. |
| Summeranter, Francis | Pvt | Hamilton Cty | 5 Feb 1846 | 96.00 | 3 Mar 1847 |
| Swyers, Daniel | Pvt | Warren Cty | 19 Apr 1816 | 96.00 | Mil. Est. |
| Taylor, Michael | Sgt | Hamilton Cty | 2 Jun 1849 | 72.00 | 13 May 1846 |
| Taylor, Samuel | Pvt | Richland Cty | 19 Apr 1815 | 96.00 | 24 Apr 1816 |
| Taylor, William | Pvt | Hamilton Cty | 1 Jul 1832 | 96.00 | Mil. Est. |
| Terleman, John | Pvt | Hamilton Cty | 10 Apr 1848 | 64.00 | Mil. Est. |
| Teypool, Ennis J. | Pvt | Hamilton Cty | 6 Jan 1847 | 96.00 | 13 May 1846 |
| Thompson, Samuel | Sgt | Sandusky Cty | 3 Feb 1816 | 96.00 | Mil. Est. |
| Thornburg, Jesse M. | Pvt | Franklin Cty | 10 Apr 1847 | 64.00 | Mil. Est. |
| Truly, Samuel | Pvt | Cuyahoga Cty | 9 Sep 1815 | 96.00 | 30 Apr 1816 |
| Truman, Charles Y. | Pvt | Muskingum Cty | 14 Jun 1848 | 96.00 | Mil. Est. |
| Turner, Edward | Pvt | Campbell Cty, KY | 18 Mar 1813 | 96.00 | 24 Apr 1816 |
| Turner, Ezekiel A. | Sgt | Lorain Cty | 29 Jul 1816 | 48.00 | Mil. Est. |
| Vanatta, Isaac | Pvt | Licking Cty | 4 Apr 1848 | 96.00 | 13 May 1846 |
| Vance, John | Pvt | Lincoln Cty | 6 Jan 1848 | 48.00 | 13 May 1846 |
| Voorhise, Luke | Pvt | Warren Cty | 1 Jan 1832 | 96.00 | 28 Jun 1836 |
| Wadkins, Darius | Sgt | Huron Cty | 14 Jun 1815 | 64.00 | Mil. Est. |
| Wallace, Nicholas | Pvt | Brown Cty | 5 Dec 1849 | 96.00 | 24 Apr 1816 |
| Wallender, Herman | Sgt | Hamilton Cty | 21 Nov 1846 | 96.00 | Mil. Est. |
| Warnock, Thomas | Pvt | Trumbull Cty | 27 Oct 1847 | 96.00 | Mil. Est. |
| Watson, Joseph | Pvt | Muskingum Cty | 6 Dec 1845 | 48.00 | Mil. Est. |
| Weber, Henry | Pvt | Hamilton Cty | 12 Jun 1848 | 96.00 | 13 May 1846 |
| Weedon, John H. | Sgt | Muskingum Cty | 27 Nov 1847 | 96.00 | Mil. Est. |
| Weenesdoefer, Lawrence | Pvt | Franklin Cty | 16 Aug 1848 | 72.00 | 13 May 1846 |
| Welch, Robert | Pvt | | 25 Jul 1821 | 96.00 | 2 Jan 1812 & 15 May 1820 |
| Whipple, Lucius | Pvt | Hamilton Cty | 30 Jun 1843 | 48.00 | Mil. Est. |
| White, Robert | Pvt | Preble, Cty | 5 May 1815 | 480.00 | 24 Apr 1816 |
| Wilks, Mills | Pvt | Boone Cty, KY | 20 Aug 1833 | 96.00 | Mil. Est. |
| Williams, John | Pvt | Ross Cty | 8 Feb 1825 | 96.00 | 24 Apr 1816 |
| Williams, John B. | Pvt | Ashtabula Cty | 15 Nov 1813 | 96.00 | 24 Apr 1816 |
| Williams, Samuel | Pvt | Holmes Cty | 12 Aug 1814 | 96.00 | Mil. Est. |
| Willis, William | Pvt | Muskingum Cty | 18 Sep 1829 | 96.00 | 24 Apr 1816 |
| Wilson, Arthur | Pvt | Tuscarawas Cty | 23 Apr 1847 | 96.00 | Mil. Est. |

| Name | Rank | Residence | Commencement of pension | Annual amount received | Act under which pension was granted |
|------|------|-----------|------------------------|----------------------|-----------------------------------|
| Wilson, James | Pvt | Holmes Cty | 18 May 1829 | 64.00 | Mil. Est. |
| Windle, Joseph | Pvt | Hamilton Cty | 20 May 1847 | 96.00 | Mil. Est. |
| Wingard, George | Pvt | Tuscarawas Cty | 29 Mar 1845 | 96.00 | Mil. Est. |
| Woods, Joshua | Pvt | Warren Cty | 22 Nov 1845 | 96.00 | Mil. Est. |
| Worcester, Lathrop L. | Corp | Hamilton Cty | 12 Apr 1848 | 64.00 | 13 May 1846 |
| Wyble, Anthony | Pvt | Huron Cty | 18 Nov 1847 | 48.00 | Mil. Est. |

# U.S. Flotilla Service

Congress created the U.S. Flotilla Service on 16 April 1814 as a separate military service.[21] The service was not a part of the navy but it was still under the control of the Secretary of the Navy. Four captains and twelve lieutenant positions were authorized for this service with the same relative rank and authority as the same grade in the navy. Captains received the same pay of a captain in the navy who was commanding a ship of 20 guns and less than 32 guns. Lieutenants received the same pay and subsistence as their counterparts in the navy. The President appointed all of the officers with the approval of the U.S. Senate.

The gunboats and barges of the Baltimore Flotilla, the Potomac Flotilla, and the New York Flotilla, operated by the U.S. Navy, were turned over to the U.S. Flotilla Service. Captain Joshua Barney took command of the Chesapeake Bay unit made up of the former Baltimore and Potomac flotilla while Captain Jacob Lewis commanded the New York unit.[22] Lewis was commissioned a captain on 26 April 1814 after having served in the navy while Barney had been an acting master commandant in the navy while the Chesapeake Bay Flotilla was being organized. On 4 March 1814 the Chesapeake Bay Flotilla had one gunboat, a pilot boat and 13 barges with 10 more barges under construction. The New York Flotilla had 38 gunboats protecting the New York harbor.[23]

In a letter to the U.S. Senate outlining the condition of the navy and the progress of the naval construction for 1814, the Secretary of the Navy William Jones said that the purpose of the U.S. Flotilla Service was to replace the officers and men in the naval flotilla service with local family men who wanted to serve in the navy but who did not want sea duty. Experienced naval officers were badly needed for the new ships being built on the east coast and for our naval squadrons on the Great Lakes. Most officers and men in the naval flotilla service wanted sea duty. Secretary Jones further stated that the new service would be governed by the same rules and regulations provided for the U.S. Navy.

The Chesapeake Bay Flotilla, under the command of Captain Barney, fought in three naval battles and one land battle during the War of 1812. The flotilla engaged the enemy's gunboats during the Battle of Cedar Point on 1 June 1814 and in the Battle of St. Leonard's Creek on 10 June 1814 and again on 26 June 1814. They were successful in tying up the British naval forces in southern Maryland during the summer months of 1814. After the final engagement, Captain Barney was forced to order the destruction of his flotilla. The men with some of the naval guns then joined the American forces defending the national capital. The Chesapeake Bay Flotilla of the U.S. Flotilla Service was the only American unit,

---

[21] *Thirteenth Congress, Session II, Chapter 59, 16 April 1814, page 125*, An act authorizing the appointment of certain officers for the flotilla service.

[22] Dudley, William S., *The Naval War of 1812, A Documentary History*, three volumes, Washington, DC: Naval Historical Center, Department of the Navy.

[23] **Condition of the Navy, and the Progress made in Providing Materials and Building Ships**, communicated to the Senate, March 18, 1814, by the Secretary of the Navy, Thirteenth Congress, Second Session, number 111, American State Papers, Naval Affairs, pp. 305-309.

which stood their grounds and fought the British during the Battle of Bladensburg on 24 August 1814. The New York Flotilla saw no action during the war.

Captain Barney was not a commodore during the war. There were no authorizations of this rank for the U.S. Flotilla Service. Commodore was a title bestowed on Barney, probably by his men, since he commanded a squadron of small vessels. It was common during this time period of American history to use the title 'commodore' if an officer commanded more than one ship, vessel or gunboat.

On 27 February 1815 the U.S. Flotilla Service was disbanded and its barges were either sold or laid up.[24] All of the men in the service were discharged and given four months extra pay.

# Fencibles, incorporated militia, and voltigeurs

While reading the histories on the War of 1812 readers are introduced to terms that are associated with the British army and the Canadian provincial troops and not associated with the American army or its militia. Fencibles, incorporated militia, and voltigeurs are just some of the terms used to describe the British and Canadian regiments and companies that fought during this war.

The American army seems to have been a lot simpler in structure than its British-Canadian counterpart. The terms "regular army," "volunteers" and "militia", and its three branches, that is, the infantry, the artillery and the cavalry, seems to be ample in identifying the divisional structure within the American army. However, the American army was just as diverse and segmented as its adversary. The American army had the same types of units within its structure as the British-Canadian army but it tended not have used the same military terms in describing these units.

First of all, there was only one American army during the War of 1812 and it encompassed all of the regular soldiers and the state and territorial militias. The army was divided into divisions and brigades and each of these components were assigned to an individual military district. Major General William Henry Harrison commanded a division of the army and this division was called the "Northwestern Army" or the "Army of the Northwest." Officially, General Harrison's command was a division and not an army.

The major difference between the American and the British armies was their regimental structure. The U.S. Army had organized an infantry regiment around ten companies of approximately 100 men each. The British, however, organized their infantry regiments around a battalion of ten companies of approximately 100 men each. Each British regiment could have up to six battalions assigned to its structure. A few regiments had more than six battalions. Thus, a British battalion was equal in strength to an American regiment and if the regiment was at full strength with six battalions or more then a British regiment was equal in size to an American division.

Eight of the ten companies in a British battalion were standard infantry while the remaining two companies were flank companies made up of light infantry and rifle companies. Light infantry were armed with muskets instead of rifles. Both types of flank companies provided the same basic support to its regiment.

The American regiment contained ten companies of infantry while the rifle and light infantry companies were organized into their own regiments. These rifle and light infantry regiments never operated as a single unit during the war as the companies were individually assigned to the various military districts in order to support the infantry in the divisions and brigades.

The British called their infantry regiments "Regiments of Foot" while the Americans used the term "Regiments of Infantry." One term used by both sides was "line regiments." Simply, these were combat regiments whether they were infantry, cavalry or artillery. The term "line" comes from the fact that these regiments were organized into lines facing an enemy before an attack was initiated.

---

[24] *Thirteenth Congress*, Session III, Chapter 62, 27 February 1815, pages 217-218, An act to repeal certain acts concerning the flotilla service, and for other purposes

"Line officers" were combat officers in line regiments. The term "line" for officers preferred to the officers' promotion system in which the officers were positioned by their ranks and dates of commission into a line or a list. This was used to determine who would be promoted next or who would receive the next open commander's position. Non-line officers were engineering, chaplain, medical and legal officers. These officers had their own officer's promotion system for their individual corps. These non-line officers would never have been given a command of a line regiment.

## The British Army

The British army was divided into the regular army and the militia. During war the militia was used to booster the strength of the army through enlistments and to be organized into "fencible" regiments until the end of a war. The British militia was never used during the War of 1812 in North America although the army relied heavily upon the Canadian militia.

During the War of 1812 Britain used regular army regiments made up of men from the British Isles, Canada and from the European continent. Two regiments were formed from Swiss, German and Dutch volunteers while two companies were organized from former French prisoners of war. There was one Canadian regiment, which was incorporated into the British army. The British also used fencible regiments from the Canadian provinces and from the West Indies.

Fencible regiments were regular army regiments that could only be used within a geographical area. Those regiments that were organized in the British Isles stayed in the islands protecting the home front and freeing the line regiments for overseas duty. Fencible regiments were organized in the Canadian provinces and many saw extensive combat during the War of 1812. Normally, fencibles were raised during a war and then disbanded once peace was obtained.

The Canadian provinces raised seven units of fencibles to serve along side of the British Army. They were organized as either infantry units or light infantry units. Six were regiments while the seventh was a company of fencibles. The West Indies provided three regiments of fencibles and a fencible ranger company for the British campaigns in Louisiana and Georgia. These units were officered by whites with black enlisted personnel.

Another type of regiment was the "invalid" regiment. These regiments were made up of pensioners, married soldiers, or soldiers receiving a disability pension. This disability would have kept a soldier from serving in a line regiment but the soldier would have to be healthy enough to be able to perform guard duty and other duties associated with maintaining a post or fort. These regiments were used heavily during the 1700's in the British Isles in order to send line regiments overseas. Two veteran battalions were sent to Canada before the war and they provided excellent service during the conflict. These battalions consisted of married soldiers who were pensioned from the army and who had re-enlisted for service in the Canadian provinces.

Artillery regiments were divided into heavy and light units. Heavy artillery was used to protect forts and fortifications while light artillery traveled with the army.

Cavalry regiments were divided into heavy cavalry, light cavalry and dragoons. Heavy cavalry, or simply cavalry, were used to attack the enemy's cavalry units or to charge the enemy's infantry lines. Light cavalry, or light dragoons, were used to skirmish behind enemy lines, disrupt communication and supply lines, provide escort duties, and to deliver dispatches for the army. Dragoons, or mounted infantry, used horses as transportation between one point to another and then the soldiers would dismount and fight on foot as light infantry. During the War of 1812 the British had only two light dragoons regiments in North America. The Canadian militia supplemented one of these regiments.

The British Army also had one fusilier regiment, which served in North America. Originally, fusilier regiments were armed with light flintlock muskets and they were used to protect artillery during a battle. By the War of 1812, the fusilier had become a fancy term for a light infantry regiment.

## The Canadian Militia

The organization of the Canadian militia was as complex as the militia of the United States. First of all there was no political entity called Canada during the war. "Canada" was a geographic name for a collection of colonies or provinces in North America administered by Great Britain. The provinces were divided into three geographic areas: Upper Canada contained the lower part of present-day Ontario, Lower Canada is present-day Quebec, and the last area was the maritime provinces of today's Canada.

The militia organization was divided by language. There was an English speaking militia and a French speaking militia. Finally, the type of militia, which served during the war, depended on how the troops were paid, either by the British government or by the local provincial government.

The Sedentary Militia was the basic peacetime militia organization made up of all able-bodied men within a province. During a war the other types of militia organization drew upon the Sedentary Militia for manpower. The Sedentary Militia was rarely called to duty during the War of 1812. When the Sedentary Militia was called to service they usually had to provide their own uniforms and equipment. The provincial government would furnish the weapons.

The Select Embodied Militia were units of volunteer or drafted militiamen organized by the provincial governments to serve as regular troops until disbanded. They were used during both war and emergencies. Upper Canada used the term "Incorporated Militia" to describe these units instead of "Select Embodied Militia." The Select Embodied Militia and the Incorporated Militia were provincial troops, which were outfitted and paid by the provincial government.

The Canadian Voltiguers was a unit of light infantry raised in Lower Canada for service during the war. The term voltigeur comes from the French Army, which raised such regiments during the Napoleonic Wars. Many of the Select Embodied Militia of Lower Canada were renamed as "Chasseur" units during the war. The term comes from the French word for "hunter" and they were also light infantry.

The Corps of Canadian Voyageurs was organized from the French Canadian fur traders who would transport military supplies and troops to the upper Great Lakes and the upper Mississippi River valley from Montreal and then return with canoes laden with furs.

The Provincial Marines were provincial troops raised as a militia naval force on the Great Lakes and Lake Champlain. The Royal Navy absorbed the unit in 1813. Another naval militia force was the Sea Fencibles, which were raised in the Province of New Brunswick during the war. They were used to protect the ports and harbors of this province.

## The American Army

The American army also consisted of a regular, full-time force and all of the militia within the states and territories. The branches of the army included the infantry, the cavalry, the artillery, the rifle, the rangers and the sea fencibles.

The infantry regiments were organized strictly as standard infantry, however, one regiment was raised late in the war as a light infantry regiment. The militia and the regular rifle regiments provided the additional support to the infantry.

Six of the 20 infantry regiments authorized in January 1813 were diverted to other purposes. The twentieth regiment was converted to ten companies of mounted rangers while five regiments, the 40th through the 44th Infantries, were raised as fencible infantry regiments to protect the coastal areas of the Atlantic seaboard and the Gulf of Mexico. These regiments could only be used to stop a British invasion from the sea. They could not be sent to Canada.

The rifle regiments, along with the artillery and cavalry regiments that were used in the war, never operated as a single unit. These regiments were organized into battalions and individual companies and assigned to all of the military districts within the United States.

The artillery regiments served both purposes of heavy and light artillery for the army. Some of the companies in the regiments were used as heavy artillery companies while others were used as light artillery companies. The light artillery regiment that operated during the war was actually used as an elite infantry regiment. The only cavalry regiments organized during the war were light dragoons regiments. The army had no regular cavalry or dragoon regiments.

The army established 17 companies of rangers during the war. These companies were mounted and the men had to supply their horses, and all of their arms, uniforms and equipment. They were given additional pay so that they could outfit themselves.

The last branch of the American army was the Sea Fencibles. These ten companies in the past have been described as volunteer militia or even as coast guard units. Army officers manned the Sea Fencibles while the enlisted men were naval seamen. This was the army's first attempt to provide costal defense protection for the country. The Sea Fencibles were an extremely unique fighting force since they could act as infantry, man artillery batteries or operate the gunboats in the harbors of our major ports.

## The American Militia

The American militia was organized under the Congressional Act of 1792 as an involuntary military organization of all able-bodied white males between 18 and 44 years of age. This act would haunt the American war efforts throughout the War of 1812. This act was 20 years out of date for it modeled the structure of the militia after the U.S. Army's organization that had been used during the Indian wars of the early 1790's. Each militia regiment was organized as a legion consisting of infantry, cavalry and artillery companies. Each regiment was designed to operate as an independent unit to put down insurrections or to operate against hostile Indians for short periods of time. The American militia was not designed to fight a European-style war.

Many of the eastern states did update their militia structure to have separate regiments of infantry, cavalry and artillery. Those militia regiments in the western area of the nation still adhered to the structure set forth by the Act of 1792 because the threat of hostile action against the Indian nations was still real.

One section of the act permitted the President of the United States to involuntary call out the militia for three months out of every year and to only use the militia within the territorial limits of the nation. Drafted militiamen would time and time again use this section of the act as an excuse not to cross over into Canada.

Six months before the war started Congress passed a series of acts for the voluntary call up militiamen to serve if war broke out between Great Britain and the United States. These acts were designed to bypass the Act of 1792. Congress failed to repeal the Act of 1792 but late in the war it did pass supplemental acts to update the original act.

Throughout the War of 1812 the President and the War Department never involuntary called out the militia thus none of the sections of the Act of 1792 were ever invoked. The problem was that the new acts set quotas on each of the states to supply men to serve with the army. Many of these states drafted militiamen in order to meet their quotas and these drafted men used the Act of 1792 to their advantage.

Like the Canadian militia, the American militia was called to serve either under federal control or under state control. However, the federal government paid for all militiamen who were called to duty.

The first act, passed in January 1812, created the U.S. Voluntary Corps, which was made up of militiamen under government control. These were the elite militia, which provided extraordinary service to the nation but they are largely forgotten today. The U.S. Voluntary Corps supplemented the army during its growing pains. The corps provided light infantry, rifle, cavalry and artillery support to the army and they were involved in nearly every battle and campaign during the war. The corps was disbanded in 1814 and the remaining men were incorporated into the army into the last four infantry regiments that were organized during the war.

The other acts for calling out the militia placed the states and territories in charge of organizing, arming and equipping regiments and brigades that were to serve a tour of duty for periods of up to six months. Very few of these militiamen ever saw combat. Many of the 6-month regiments from New York, Pennsylvania, Ohio, Kentucky, Louisiana, and Tennessee were used in combat and they performed well.

Most of the 3-month militiamen were used to man forts and fortifications, to build roads and facilities, to deliver supplies and dispatches, and to perform other duties that would free up the regular army so that they could be used for combat.

The majority of 6-month militia were organized into infantry regiments of about 500 men, thus comes the statement that the militia regiments during the war were at half strength. They also provided light infantry, rifle, dragoons and artillery support to the regular army to a lesser degree. Most state and territorial militias could not finance the outfitting of light dragoons and artillery companies. Some of the states on the eastern seaboard organized their own sea fencible companies.

The last area of support that the militia supplied to the army was ranger support. Many militia ranger companies were called to duty in the west and they provided outstanding service to the country. Some of the states and territories called their rangers "spies."

## The Legion

Another military term that was used on both sides of the war was "legion." A legion was a military unit made up of companies from more that one branch of an army.

The Army of the Northwest had two brigades of infantry and a legion as part of its organization. The legion contained all of companies of light dragoons, artillery, rifle and the U.S. Volunteers that were assigned to this division. The Prince Edward Island militia in the Maritime Provinces organized a legion during the later part of the war and the Territory of Michigan had a legion before this territory fell to the British.

## The Engineers

Each side had a corps of engineers organized into companies of sappers and miners. Sappers were combat engineers who were used in the construction of military buildings and fortifications. They were also used to repair military facilities. Miners were used to dig tunnels under enemy fortifications in order to set off charges to destroy such facilities.

A certain number of soldiers in each infantry regiment were designated as "pioneers." These were soldiers who built roads, dug trenches and constructed bridges as an army advanced. In the western areas of the conflict militia companies performed most of these duties.

## Militia, U.S. Volunteers, National Guard:
## What's the difference?

Three military organizations which popped up during the 1700's and 1800's, causing great confusion when researching an ancestor who served in the military, were the militia, the U.S. Volunteers and the National Guard. The militia did not become the U.S. Volunteers and both types of military organizations did not become the National Guard. Each of these military organizations existed separately to supplement and to support the U.S. Army during wartime. The major differences were how they were controlled and how they were funded.

The militia has existed since the founding of the first British colony in 1607 at Jamestown, Virginia. The militia is a peace-time military organization in which all able-bodied men are required to be a member. These militiamen, by law, must provide their own weapons, uniforms and equipment. Under the Militia Acts of 1792 and 1795, militia units were controlled by the governor of each state or territory, and they could be used to defend a state or a territory.

The militia could also be activated to serve with the U.S. Army in times of war. In most cases, the militia that was called up during a conflict, formed new militia regiments, separate from the peace-time regiments. These regiments were made up of militiamen who had volunteered to serve or by men who had been drafted. Militia officers received a state officer's commission. The state militia regiments, which supplement the U.S. Army during a war, were still under control of their governors.

Officer's positions in the peace-time militia were actually political positions in which elections were held to fill the officer's ranks. Officers served a term of office for five years. If a man was elected as captain, he had to re-run for that office after five years if he wanted to keep his captain's rank. If he wanted to be promoted to major, then he ran for a vacant major position when a position was open and an election was held.

Militia officers who were called to duty during war-time and who received a higher officer's rank, held this rank only for the duration of their tour of duty. Once their tour of duties were over, these officers went back to their peace time ranks. This was the same condition for militia officers who received officer's positions in the U.S. Volunteers. Both federal and state officer commissions for militia officers were temporary wartime commissions.

The U.S. Volunteers was a branch of the U.S. Army which was activated only during a war. They are members of the U.S. Army but were short-termed soldiers who were discharged at the end of each conflict. They were used to bring the regular army up to full strength and also to increase the size of the army during wartime.

A volunteer was also one who voluntarily served a tour of duty with a state militia regiment or enlisted either in the regular army or a U.S. Volunteer regiment. Many militiamen were drafted and that is why some militia regiments are referred to as volunteer militia regiments and other militia regiments as drafted militia regiments.

U.S. Volunteers were armed, equipped and clothed by the U.S. Army. The officers received a federal officer's commission from the President of the United States and these officers were equal in all respects to regular army officers. During the Revolutionary War and during the Indian Wars of the 1790's, the volunteers were called "levies." During the War of 1812, they were officially called the U.S. Voluntary Corps.

The states raised U.S. Volunteer regiments during the Mexican-American War, the Civil War and the Spanish-American War. The federal government raised U.S. Volunteer regiments during the War of 1812, a few regiments during the Spanish-American War, and all of the Volunteer regiments during the Philippine-American War. When the state-raised U.S. Volunteer regiments were federalized, the states no longer had control over these regiments.

During the Civil War, most of the state militias which were called for military duty were federalized and the officers received federal officer's commissions, thus they became U.S. Volunteers. However, many state militias were called to duty by the federal government in order to protect their states but they were never federalized. These troops were normally called up during a temporary crisis and served only a month. The officers retained their militia officer's commissions.

By the time of the Civil War, most of the state militias no longer required all able-bodied men to be members. The militias were very small in numbers but they still required their men to supply their own weapons, uniforms and equipment.

Towards the end of the Civil War, the federal government, in cooperation with the states and territories, created the National Guard. The National Guard replaced the militia. The militia did not become the National Guard. The militia still exists today in twenty-two states as it will be explained later in this article.

The National Guard was funded by both the states and the federal government and they received the standard weapons, uniforms and equipment as used in the U.S. Army. In theory, when a National Guard unit was activated for federal duty, they were already armed, uniformed and equipped, and thus ready for duty.

Men of military age enlisted in the National Guard. It was not required for all men to be a member of this organization. Towards the end of the Civil War, National Guard battalions were called up to form U.S. Volunteer regiments. These regiments were federalized and their officers received federal officer's commissions.

During the Spanish-American War and the Philippine-American War, all of the National Guard units which were activated, were formed into U.S. Volunteer regiments and their officers received federal officer's commissions. Those men who enlisted during these wars, served in either the U.S. Army regiments or in the U.S. Volunteer regiments.

The Dick Art of 1903 reorganized the U.S. Army and it made the National Guard a member of the U.S. Army replacing both the U.S. Volunteers and the militia. This act replaced the Militia Acts of 1792 and 1795. The National Guard units were numbered by the federal government and when they were federalized, they became a part of the U.S. Army.

During World War I, the first twenty-five infantry divisions were U.S. Army divisions while the second twenty-five divisions were National Guard divisions. Divisions that started with number 51 as their designations were "National Army" divisions made up of men who enlisted or who were drafted.

As stated, the militia still exists today. There are now called the State Guards, the State Military Reserves or still, the State Militia. Collectively, they are called the State Defense Forces and twenty-two states have these organizations. For the most part, these militia units are made up of volunteers who supply their own uniforms and equipment. Most are unarmed. There are used by the states when the National Guard is activated for federal duty.

The State Defense Forces have the same mission as the National Guard does during peacetime, that is, to protect the state and to assist the state in time of crisis and disasters. A few of the states have naval and air militias. Ohio's State Defense Force, which is under the control of the Adjutant General's Department, is made up of the Ohio Military Reserves (the militia) and the Ohio Naval Militia.

In short, the militia was a peace-time military organization which was used to supplement the U.S. Army during wartime. The U.S. Volunteers was a branch of the U.S. Army which was only activated during wartime. The National Guard was the best of both worlds in which it was a peacetime military organization which in theory was ready for action when called to federal duty.

## Anatomy of a militia company:
## The myths, the misconceptions and the draft [25]

One of the biggest misconceptions of the War of 1812 is that the men from the same community or general area of the state banded together to form companies, elected their own officers, and then marched off to fight the British. This did not happen! The only time in American history that this occurred was at the beginning of the Revolutionary War when the British Crown still controlled the colonial militia and the rebel forces (the good guys) organized the original state militias.

Prior to 1776, the rebel areas of New England did organize new militia companies and elected their own officers. Later, the rest of the colonies did the same thing. These were the minutemen companies. As the British Crown lost control of the colonies, the new rebel governments created the state militias and then they passed laws to control their militia forces.

The colonial militias (loyalists) under the Royal colonial governors did not elect their officers. These officers were appointed by the governors and they were usually the elite gentlemen of a community who were loyal to the king.

Also, men living in the unorganized areas of our nation, throughout our history, did organize into companies and elected their officers. These ad hoc companies were usually organized to fight the Indians. Most of these companies were made up of squatters who settled on government lands without permission.

The call up of Ohio's militia troops for the War of 1812 was not a democratic process. It was predetermined by state law on how the officers and men were selected to serve during this war. This process for calling up the militia was called the 'draft.'

Individual men were not drafted to serve in our military until the Civil War. The concept of the draft in 1812 is totally different than the drafts of the late nineteenth and twentieth centuries. During the War of 1812 groups of militiamen, regardless of their names, were drafted, not the individual man. During and after the Civil War, a lottery was held and individual men's names where drawn to serve in the army, regardless whether that person was a militiaman or a civilian.

Regimental and brigade officers were selected by seniority throughout the war. If the governor tasked the First Division of the Ohio Militia to raise a brigade for a tour of duty then the senior brigadier general in the First Division was selected to command this brigade. On the second call up for a brigade from the same division, then the brigadier general next in line on the seniority list would command this brigade.

Colonels were selected in the same way for command of regiments. During the first half of the War of 1812, brigades were raised from the same militia divisions. The second half of the war saw one regiment being raised in the division where the brigadier general lived while the second regiment was raised from another militia division in the state.

The makeup of a brigade varied throughout the war. Normally, there were two regiments, each with 500 men, and any number of spy (ranger), artillery, cavalry or rifle companies as needed for the tour of tour. Some brigades had three regiments; others had two regiments and a battalion.

The structure of a militia company during the War of 1812 was identical to the structure of the militia companies formed during the Revolutionary War. The number of men in a company would vary over time but the structure remained the same.

Companies were made up of a captain, a lieutenant, an ensign, four sergeants, four corporals, two musicians and 64 privates (77 men). By Ohio militia law, a company could have as few as 40 privates and as many as 100 privates under certain conditions.

The 64 privates were divided into eight classes of eight men each. The classes were numbered from one to eight. The youngest men were normally in the lower numbered classes while the older men were in the higher numbered classes. The class was the smallest unit in the militia. The eight men of each class messed together and slept together, and they were drafted together. Two classes made up a squad while four classes made up a platoon.

---

[25] This article is based upon the militia laws of the State of Ohio during the War of 1812 and some items may defer in other states and territories.

Militia infantry companies were rarely called up to serve together as a militia company during the war. During a call up for duty, a regiment may have been tasked to raise one or two companies. If tasked for two companies, then each battalion in the regiment would raise a company. Regiments had eight companies, organized into two battalions of four companies each.

If the regiment was tasked for only one company then the 1$^{st}$ Class in each of the eight companies of the regiment were drafted to serve in this call up. During the second call up for troops, then the 2$^{nd}$ Class from each of the eight companies in the regiment were drafted. This gave the new companies 64 privates.

After the classes were drafted, then the men in the effective classes could obtain a substitute to serve for him. There were no provisions in the militia laws on how a substitute would be compensated for his services. The substitute could have voluntarily served for the draftee or the draftee may have paid him to serve. Regardless, the substitute was paid his military salary and only he qualified for land bounties and pensions after the war. The draftee who obtained a substitute received no pay or benefits.

When an artillery, spy, cavalry, light infantry, or rifle company was tasked to serve then half the men were drafted while the other half stayed home. These companies were then at half strength during their tour of duties. Some half strength companies merged to form a full company.

Company officers were selected for a tour of duty by their seniority through a chart that was determined by Ohio's militia laws.

| Company | Captains | Lieutenants | Ensigns |
|---|---|---|---|
| 1$^{st}$ draft | 1$^{st}$ | 2$^{nd}$ | 4$^{th}$ |
| 2$^{nd}$ draft | 2$^{nd}$ | 1$^{st}$ | 3$^{rd}$ |
| 3$^{rd}$ draft | 3$^{rd}$ | 4$^{th}$ | 2$^{nd}$ |
| 4$^{th}$ draft | 4$^{th}$ | 3$^{rd}$ | 1$^{st}$ |
| 5$^{th}$ draft | 5$^{th}$ | 6$^{th}$ | 8$^{th}$ |
| 6$^{th}$ draft | 6$^{th}$ | 5$^{th}$ | 7$^{th}$ |
| 7$^{th}$ draft | 7$^{th}$ | 8$^{th}$ | 6$^{th}$ |
| 8$^{th}$ draft | 8$^{th}$ | 7$^{th}$ | 5$^{th}$ |

Based upon the above chart, during the first draft for a new company, the senior captain in the regiment was selected as the commander of this company. The lieutenant with the second oldest seniority was selected for duty while the ensign with the fourth oldest seniority was also selected. During the second call up for troops, the captain with the second oldest seniority was selected to command the new company, and then the lieutenant and ensign were selected accordingly.

An election of officers did occurred once the new companies arrived at their rendezvous locations and they were organized into battalions and regiments. Each battalion needed one major and there were officer's positions needed to be filled on the regimental and brigade staffs. The officer corps would then elect officers for these positions, and then the companies would hold elections to replace the missing officers in their companies.

Provisional companies were also formed during the War of 1812 and this too was controlled by Ohio's militia laws. In the frontier areas of Ohio, the militia companies were exempt from being called up and for serving in another part of the state. Their sole duty was to build blockhouses for the protection of their family and friends, and if need be, to evacuate these civilians to the safer areas of the state.

These provisional companies were not activated to serve in the war and these men were not paid by the federal government for the services. You probably will never find a muster roster for these companies.

## Wartime ranks

Nearly all wartime military ranks and promotions were temporary. There were militia captains who were called up for a tour of duty and elected to either a major's or to a colonel's position. Once their tour of duty was completed and they returned home, they went back to their captain's ranks. There were no wartime commissions for militia officers. They were appointed or elected to a wartime position and served in this capacity only during their tours of duty.

Officers, up to the rank of brigadier generals, were elected for a five-year term either by their peers or by their company during peace-time. Major generals were selected by the governor and approved by the Ohio Assembly.

Militia officers who were selected for the U.S. Voluntary Corps received a one-year commission from the president of the United States. The U.S. Volunteers lasted for two years and many officers had their commissions renewed. After two years, the officers and men who still had enlistment time were transferred to the 45th through 48th Regiments of Infantry, and the officer's commissions then had to be approved by the U.S. Senate.

With the expansion of the regular army throughout the war, the army selected many militia officers to hold officer's positions in the new army regiments. These officers received their regular army officer's commissions with the approval of the U.S. Senate. Many men resigned their militia commissions at this time.

As an example: John Campbell was a lieutenant colonel in a regiment located in Portage County. He accepted a commission as a captain in the U.S. Voluntary Corps. He was still a lieutenant colonel in the Ohio militia. His second-in-command filled in for him until his released from the corps. After his return to his county and his service ended with the U.S. Volunteers, Campbell would be elected as a brigadier general and he would command his county's militia brigade.

The best example for multiple commissions involves Robert Lucas of Portsmouth, Ohio. He was the brigadier general for the brigade located in Scioto County and he raised an infantry company for Brigadier General William Hull's Army of the Northwest in 1812. He enlisted in this company as a private. He never resigned his militia generalship.

When the army was being organized in Urbana, Ohio, he was recognized and commissioned as a captain in the U.S. Voluntary Corps. He then served as a scout for this army. He was still a general in the Ohio militia. When the army reached Detroit, he received a dispatch informing him that he had received a commission as a captain in the U.S. Army.

General Hull surrendered his army to the British on 16 August 1812. Lucas became a prisoner of war on parole status, he returned home, lost his captaincy with the corps and resigned from the U.S. Army (having never served).

Later in the war, he received another commission in the U.S. Voluntary Corps as a lieutenant colonel with the responsibility of commanding all of the U.S. Ranger companies in the Old Northwest Territory. He also turned down this commission. After the war, Lucas was promoted to major general in the Ohio militia, and then as the commander in chief when he became the 12th governor of Ohio.

## Volunteer militia versus drafted militia

The federal Militia Act of 1792 gave the president of the United States the power to call up the militia involuntary for up to a three-month period, once a year, when the country was in endanger of an Indian or foreign power invasion. These militiamen who served a tour of duty under these circumstances were called 'drafted militia.'

The militiamen who volunteered to serve for a three-month tour or a six-month tour were called 'volunteers.' The U.S. Voluntary Corps was a wartime branch of the U.S. Army. The men who served in this corps were either commissioned by the president or enlisted for one year. These men were called 'U.S. volunteers."

The U.S. Army did not draft men during the War of 1812. All of the men who were commissioned or who enlisted in the army were volunteers.

## Drummer Boys and fifers

One of the biggest myths associated with the early wars of this nation is that all drummers and fifers were under-aged boys. This myth has been perpetuated by Hollywood in scores of films for over the past 100 years. In reality, under-aged boys were in the minority when serving as musicians.

The Ohio militia did not keep records of a militiaman's age, birth or civilian occupation but the U.S. Army did keep these records. From U.S. Army enlistment records for the War of 1812, 69 Ohioans enlisted as musicians in this war and only 12 were under the age of 18. Thus, 17% of the musicians were under-aged. The youngest musician was 13 years old while the oldest was 44. The majority of the men were in their twenties.

The percentage of under-aged boys serving as musicians in the militia may be greater than the percentage that served in the U.S. Army. With shorter tours of duty and with most men serving within their state, there may have been more under-aged boys serving in their father's militia companies.

## Lack of weapons

Another misconception is that many of the militiamen did not own firearms and that is why these men reported for duty without muskets or rifles. In reality, most men did own a weapon but these weapons were of a smaller caliber that was designed for shooting birds and small game, and not men. These weapons were left behind in the care of teenage sons so that they could continue to provide food for the family tables.

Many men brought to the battlefield the muskets that their fathers had used during the revolution. Many of these muskets were of the proper caliber but they were old and worn, and many were in need of repairs. Other men simply showed up for their tour of duty without firearms in hopes of being issued a new musket or rifle, which they assumed that they would be able to keep once their tours of duty were over.

## Officer's background

Officers in the state militia tended to be the leading citizens of their townships or their counties. The majority of these officers from the rank of captain through major general had been citizens of Ohio for many years and had served in the state militia since either becoming of age or taking up residency. Officers tended to be landowners, better educated and had previous military experiences either with the militia or with the regular army.

The senior officers, that is, the majors, colonels and generals, tended to be professionals, lawyers, public servants, or newspaper publishers. Many of the senior officers had served during the Indian Wars of the 1790's either in the regular army or with one of the state militias.

Most captains were in their 30's and early forties. The junior officers, that is, lieutenants and ensigns, tended to be the sons of the captains, staff officers and generals. Many of the older men had Revolutionary War experiences, but this was the exception not the norm.

## 1st, 2nd, 3rd and 4th sergeants, etc.

Many muster rolls list the sergeants and corporals as the 1st sergeant, 2nd sergeant, 3rd sergeant and 4th sergeant, while corporals had the same designations. These designations were not ranks but positions of seniority. There were only two sergeant ranks in the Ohio militia. The sergeant major was the senior enlisted men who reported directly to the regiment's colonel. Each regiment had one sergeant major's position. Each company had four sergeants and four corporals. All of the company sergeants were paid the same as were the corporals for their rank. The sergeant major had the highest salary of all of the enlisted men.

When marching, the 1st sergeant, 1st corporal and the musicians marched with the captain. The 2nd and 3rd sergeants and the 2nd and 3rd corporals marched with the lieutenant while the 4th sergeant and the 4th corporal marched with the ensign.

Militia artillery and cavalry companies had two lieutenant's positions and no ensign positions. Some musters designated these lieutenants as 1st lieutenant and 2nd lieutenant. The same holds true with the lieutenants as with the sergeants and corporals, these designations were positions of seniority and not an actual rank. The regular army did have the ranks of first lieutenant, second lieutenant, and third lieutenant, with each rank having its own pay grade.

The 1st through 4th designations were eliminated in this book since they are not relevant to the ranks of sergeant and corporal. The rank of first sergeant (the senior sergeant in the company) was not created by the army or the militia until after the War of 1812.

## Lieutenants commanding companies

There are many muster rosters of companies that were commanded by lieutenants, ensigns, and even sergeants, but this is not accurate. Only captains commanded companies. If there was a vacancy for the captain's rank, the company would have elected a new captain before the regiment had marched to the front.

The lieutenants, ensigns and sergeants on these rosters are actually commanding 'detachments' and not companies. There are captains who were tasked to detach part of their companies to perform certain duties and another officer or non-commissioned officer in the company was selected to command there 'detachments.' Once the task was completed, the 'detachment' would have merged back into its parent company.

Only the U.S. Army had lieutenants, normally first lieutenants, commanding companies during this war under certain circumstances. If a captain was selected to hold an officer's position on the brigade or divisional level, the first lieutenant was given temporary command of the company. Also, if the captain had been wounded or recovering from sickness, then the lieutenant would be in charge.

If a captain had been killed, had resigned his commission or had been forced out of the service due to a court-martial, then the lieutenant would be in command until another captain was selected for the company. The army promoted on the seniority system, so a lieutenant commanding a company did not always receive a promotion to captain to fill the vacancy in his company.

## Types of companies

Many of the infantry companies on muster rolls are mislabeled for they were actually mounted infantry companies or dragoon companies. You will find cornets, saddlers and furriers listed in some of these infantry companies. These ranks are unique to dragoon and cavalry companies and not to infantry companies.

Mounted infantry companies are not true cavalry units. They were foot soldiers that used horses for transportation. Once on the battlefield, the men dismounted and fought on foot. Another term for 'mounted infantry' was 'dragoons.'

Light dragoons were lightly armed cavalry companies whose mission was to disrupt the enemy behind their lines by raiding supply lines and to force the enemy into minor skirmishes. Most of the light dragoons were used as escorts for generals and as dispatch riders during the war. By law, Ohio's light dragoons companies were called 'cavalry' companies. Some of Ohio's cavalry companies were called 'light horse' companies.

During this period of time where were three types of cavalry units. The traditional cavalry, the light dragoons and the mounted infantry (or dragoons). These three cavalry-type units had three different missions, used a different type of horse and were armed differently. The traditional cavalry companies were not used by either the British or the Americans during the War of 1812. The U.S. Army would not have true cavalry regiments until the Civil War.

Spy companies were used as reconnaissance units. These companies are actually ranger companies. Spies were sent behind enemy positions in order to gather information on the enemy. These companies had a different mission than the cavalry and infantry units, and depending on a particular mission the spies could either be mounted or on foot.

Besides infantry, mounted infantry (dragoons), light dragoons and spies (ranger) companies, the militia also had artillery, rifle and light infantry companies. Artillery companies in the militia were rare in the western part of the country because of the costs of outfitting such a company. Many states and territories created artillery companies only after they had received field pieces from the federal government.

Rifle companies and light infantry companies had the same mission but the rifle companies were armed with rifles instead of musters. These companies patrolled the areas in front and on the sides of regiments or brigades then they were marching. They were also used to flank the enemy during a battle and they were normally the last companies to leave a battlefield.

# U.S. Revenue Marine

The U.S. Revenue Marine was the first naval service organized under the Constitution of the United States. Congress authorized the construction of ten cutters under the Tariff Act of 4 August 1790 with a crew of ten men each.[26] The primary duties of this new service were to enforce the laws governing the collection of customs and to stop smuggling & piracy along our shores.

The original act stipulated that the master of each cutter would have the equivalent rank of a captain in the army. Since the navy had not been created yet, the rank, pay and subsistence of each member of the crew would parallel the army's rank structure. The three mates had the subsistence of an army's lieutenant, while the four marines and the two boys had the subsistence of a soldier in the army.

The service would also be known as the Revenue Service and on 31 August 1894 it would officially be renamed the U.S. Revenue Cutter Service. On 28 January 1915 the service would merge with four other services to form the U.S. Coast Guard. During the War of 1812 the Revenue Marine was under the control of the Treasury Department and the service took an active role in fighting the British alongside the U.S. Navy.

The President had the power to appoint the officers of each cutter who would also serve as custom officers. The vessels of this service would be called revenue cutters regardless of the type and size of the sailing ships.

The Act of 18 April 1814 provided a pension for those who had been wounded or disabled while cooperating with the navy during the war.[27] These men were placed on the navy's pension list and they received the same compensation by rank as the regular navy.

The U.S. Revenue Marine captured seven British warships and privateers during the War of 1812 and in turn lost four cutters to the British and one to a fire.[28] The service operated in the coastal waters of the United States.

---

[26] *First Congress*, Session II, Chapter 35, 4 August 1790, pages 145-178, An act to provide more effectually for the collection of the duties imposed by law on goods, wares and merchandise imported into the United States, and on the tonnage of ships or vessels.

[27] *Thirteenth Congress*, Session II, Chapter 65, 18 April 1814, page 127, An act granting pensions to the officers and seamen serving on board the revenue cutters in certain cases.

[28] Silverstone, Paul H., *The Sailing Navy, 1775-1854*, (Annapolis, Maryland: Naval Institute Press, 2001), chapter 4, United States Revenue Cutter Service.

## Ohio's Second Recruiting District

Lieutenant Colonel George Tod of Youngstown, Ohio, has the distinction of serving as a senior recruiter during the War of 1812 in three different infantry regiments: the 17th Regiment of Infantry, the 19th Regiment of Infantry and in the reorganized 17th Regiment of Infantry. Before becoming an officer in the United States Army, Tod was the brigade major and inspector for the 3rd Brigade of the 4th Division, Ohio Militia, which was under the command of Brigadier General Simon Perkins.[29]

The 17th Regiment of Infantry was organized under the Congressional Act of 11 January 1812.[30] On paper this regiment consisted of two battalions of nine 110-man companies each.[31] One battalion was headquartered in Kentucky while the other battalion was headquartered in Chillicothe, Ohio.

**Ohio in 1812**

**2nd Recruiting District**

"District of Zanesville"

The first two months of 1812 were spent recruiting the officers for this regiment. It wasn't until 12 March 1812 that these officers were commissioned into the army and the task of forming the rest of the regiment could begin. On 28 April 1812 Brigadier General James Winchester, commander of the Army of the Northwest, ordered Major Tod to Lexington, Kentucky, for the purpose of receiving the monies and the necessary papers in order to form a recruiting service in Ohio.[32] His official orders were dated 9 May 1812.[33]

The *Historical Register and Dictionary of the United States Army From Its Organization, September 29, 1789, to March 2, 1903* shows that Major Tod was commissioned in the 19th Regiment of U.S. Infantry on 12 March 1812.[34] The 19th Infantry would not be authorized until 26 June 1812 and it would not be until July before any officers would be assigned to this regiment.[35] Major Todd was assigned to the 17th Infantry on 12 March 1812 and he was transferred to the 19th Infantry on 6 July 1812.

---

[29] *Western Reserve and Northern Ohio Historical Society*, Tract Number 15, April 1873, (Cleveland, Ohio: 1877), page 1, Letter from Brigadier General Simon Perkins to Major George Tod, 27 April 1812.

[30] Heitman, Francis B., *Historical Register and Dictionary of the United States Army From Its Organization, September 29, 1789, to March 2, 1903*, Volume I, (Genealogical Publishing Company, Baltimore, Maryland: 1994), page 113, Seventeenth Regiment.

[31] Mahon, John K. and Romana Danysh, *Infantry Part 1: Regular Army*, Army Lineage Series, (Office of the Chief of Military History, United States Army, Washington, DC: 1972), page 14, History of the Organization of the Infantry: *Through the Second War with England.*

[32] *Western Reserve*, Tract Number 15, April 1873, (Cleveland, Ohio: 1877), page 2, Letter from Brigadier General James Winchester to Major George Tod, 28 April 1812.

[33] *Ibid.*, Tract Number 17, November 1873, (Cleveland, Ohio: 1877), page 1, Letter from Brigadier General James Winchester to Major George Tod, 9 May 1812.

[34] Heitman, *Historical Register,* Volume I, (Genealogical Publishing Company, Baltimore, Maryland: 1994), part II, complete alphabetical list of commissioned officer of the Army between 29 September 1789 and 2 March 1803, pp. 149-1069.

[35] *Ibid,*, page 116, Nineteenth Regiment.

Recruiting instructions from the War Department had been sent to the commanders of all of the army's departments early in 1812.[36] These instructions formed a recruiting service for the army, which divided the country into recruiting districts. Although each regiment recruited their own men, the Adjutant General's Department of the War Department maintained control on how the men were recruited. The department also gave each regiment instructions outlining the duties and responsibilities for each recruiting officer.

Enclosed with the instructions was a paragraph entitled "To Men of Patriotism, Courage and Enterprize," which was to be used for newspaper ads and billboards to help the recruiting efforts. A sample recruiting contract was also included.

Ohio was divided into two recruiting districts with Chillicothe serving as the headquarters for the first district and Zanesville as the headquarters for the second district. Both of these cities would serve as rendezvous points for the new recruits. Later, all recruits would be sent to Chillicothe to be formed into companies made up of men from all over Ohio.

After the two districts became operational, Major Tod took command of the 2nd Recruiting District, also called the District of Zanesville, for the 17th Infantry.[37] The following counties were included in this district:

2nd Recruiting District

| County | Civilian Population |
|---|---|
| Cuyahoga & Huron | 1,459 |
| Ashtabula & Geagua | 2,000 |
| Trumbull | 8,671 |
| Portage | 2,995 |
| Columbiana | 10,878 |
| Stark & Wayne | 2,734 |
| Coshocton & Tuscarawas | 3,045 |
| Knox | 2,149 |
| Licking | 3,852 |
| Muskingum | 10,036 |
| Guernsey | 3,051 |
| Belmont | 11,097 |
| Jefferson | 17,260 |
| Washington | 6,000 |

Most of these counties in 1812 were far larger in area than their current bounties. These counties represent about half of the area of Ohio in 1812.

The following officers from the 17th Infantry and the 2nd Regiment of U.S. Artillery were assigned to the 2nd Recruiting District:

Captain Wilson Elliott
Captain Abraham Edwards
Captain James Herron
1st Lieutenant Joseph H. Larwill, Artillery
1st Lieutenant Samuel Booker

---

[36] "Recruiting Instructions", Adjutant and Inspector General's Office, Washington, DC, 1812, *War of 1812 Collection*, Appointments, Resolutions and Certificates, Western Reserve Historical Society Archives Library, Cleveland, Ohio, manuscript section, call number MS-660, container 1, folder 10.

[37] "2nd Recruiting District, 17th Regiment of Infantry", 14 June 1812, *George Tod Collection*, Western Reserve Historical Society Archives Library, Cleveland, Ohio, manuscript section, call number MS-3202, container 2A.

1<sup>st</sup> Lieutenant George W. Jackson
2<sup>nd</sup> Lieutenant Charles Esti, Artillery
2<sup>nd</sup> Lieutenant Henry Frederick
2<sup>nd</sup> Lieutenant Robert Morrison
Ensign John Milligan
Ensign Batteal Harrison

The District of Zanesville was further divided into three sections based upon the civilian population.[38] The northern section included the counties of Trumbull, Ashtabula, Geagua, Cuyahoga, Portage, Wayne, Stark and Columbiana. The total population of this area was 28,749.

The southern section included the counties of Muskingum, Washington, Guernsey, Coshocton, Licking, Knox and Tuscarawas. The total population for the southern section was 28,133. The eastern section included Jefferson and Belmont Counties, totaling 28,367 civilians.

Captain James Herron, Lieutenant George W. Jackson, Lieutenant Robert Morrison, and Ensign Batteal Harrison were assigned to the northern section. Captain Wilson Elliott, Lieutenant Samuel Booker, Lieutenant Henry Frederick and Ensign John Milligan were assigned to the southern section. Lieutenant Joseph H. Larwill was assigned to the eastern section while Lieutenant Charles Esti was assigned to Dayton, Ohio.

The names of Captain Edwards and Lieutenant Esti do not appear on any muster or returns for this recruiting district. It is doubtful that these men served as recruiting officers. The purpose of Lieutenant Esti being assigned to Dayton in the western part of the state is not known.

Captain Edwards, although commissioned as an officer in the United State Army, never served in the 17<sup>th</sup> Infantry. When his commission was approved he was already serving as the chief surgeon for Brigadier General William Hull's Army of the Northwest. He would later serve as a deputy quartermaster general in the army between 15 March 1814 and 15 June 1815 holding the rank of major.

Lieutenant Esti's name does not appear in the *Historical Register and Dictionary of the United States Army From Its Organization, September 29, 1789, to March 2, 1903* for officers who were commissioned in the United States Army. His name may have been missed when this book was being prepared or Congress never approved his commission.

Major Tod's recruiting tasks included the assignment of officers and enlisted personnel to recruiting stations within the state. When recruiting fell off at a station it was his responsibility to reassign his men to another station. He was the paymaster whose duties included the payroll and voucher payments. He was also the quartermaster who was in charge of outfitting the recruits with their initial clothing and equipment. Besides supplying men for the 17th Infantry, Major Tod was to assist with the recruiting efforts of the 2nd Regiment of Artillery, which was forming a company from within Ohio.

Weekly recruiting reports were required to be submitted to headquarters by each of the stations. Once 100 men had been recruited, the men would be ordered to "rendezvous" at Zanesville where they would be organized into a new company. Later in the war Chillicothe was selected as the rendezvous point.

Many times upper command would appoint some of the recruiting officers for officer positions within a new company. It was Major Tod's duty to fill the remaining officers' slots and also to ensure that all of the non-commissioned officers had been assigned.

Writing from Zanesville to General Winchester on 29 June 1812, Major Tod stated that he had contracted a building to serve as a barracks that could house 100 men and that he had obtained the necessary supplies and rations.[39] He had employed a physician but he did not have any junior officers to fill the positions of assistant quartermaster or acting adjutant. He also had 200 sets of summer uniforms in stock that would be issued to the new recruits.

[38] The division of the 2<sup>nd</sup> Recruiting District", *George Tod Collection*, Western Reserve Historical Society Archives Library, Cleveland, Ohio, manuscript section, call number MS-3203, container 2, folder 6.

[39] *Western Reserve*, Tract Number 15, April 1873, (Cleveland, Ohio: 1877), pp. 2-3, Letter to Brigadier General James Winchester from Major George Tod, 29 June 1812.

Recruiting throughout the state was extremely slow until the declaration of war on 18 June 1812. Men were reluctant join the army during peacetime and once war came the men would rather serve in the militia than the regular army because of the shorter enlisted times offered by the militia. Recruiting did increase after the surrender of General Hull's army to the British on 16 August 1812 at Detroit.

On 26 June 1812 the army reorganized its regimental structures and standardized the manpower strengths for each type of regiment.[40] The 17th Infantry was downsized into ten 106-man companies with an authorized manning of 1,070 men. At the same time, the second battalion was split off to become 19th Regiment of Infantry with the same structure and manning as the 17th Infantry.[41]

Although the recruiting service in Ohio was fully functioning by June it appears that no company of Ohioans was activated for the 2nd battalion of the 17th Infantry. The recruiting service in Ohio, with all of the officers and men plus the new recruits, was transferred to the 19th Regiment of U.S. Infantry. Major Tod would spend the next two years as the senior recruiter for the 19th Infantry.

By July 1812 recruiting stations for 2nd Recruiting District had been set up in the following Ohio cities and villages with an officer and a sergeant plus a musician assigned to each station. Normally, there was one officer assigned to the recruiting duties while the sergeant's responsibility was to give each recruit his initial military training.

July 1812

| Station | Recruiter |
| --- | --- |
| St. Clairsville | Lt. Samuel P. Booker |
| Warren | Capt. Wilson Elliott |
| Cadiz | Ensign James Milligan |
| New Lisbon | Lt. Henry Fredericks |
| Canton | Lt. Joseph H. Larwill |
| Chillicothe | Capt. James Herron |

On 29 July 1812 Colonel John Miller, commander of the 19th Infantry, ordered Major Tod to form the recruits into a new company at Zanesville.[42] This company was scheduled to meet a detachment from the 17th Infantry and then march to Detroit as reinforcements for the Army of the Northwest. Colonel Miller was at his headquarters in Chillicothe when he wrote this letter.

Once a company was formed it was then ordered to Franklinton (now Columbus), Ohio, were it was issued its field equipment and any other required items that had not been supplied at "rendezvous" or at the recruiting stations. The company would then report to the regiment in the field.

Surprisingly, most recruits during the war were not used to back fill the open positions in the existing companies. As attrition took its toll on a company through deaths, desertions and enlistment separations, the size of a company would dwindle down. Once this happened, the enlisted men were reassigned to another company and the officers and sergeants were used to backfill vacancies in other companies or they were sent home to the recruiting service.

By August 1812 Lieutenant Larwill had relocated his recruiting station from Canton to Cadiz, Ohio, and Ensign Harrison was assigned the recruiting duties at Cambridge, Ohio.

---

[40] Heitman, *Historical Register*, Volume II, (Genealogical Publishing Company, Baltimore, Maryland: 1994), pp. 572-573, Organization of the Army under the Act of June 26, 1812.

[41] *Ibid.*, Volume I, page 116, Nineteenth Regiment.

[42] *Western Reserve*, Tract Number 15, April 1873, (Cleveland, Ohio: 1877), page 3, Letter from Lieutenant Colonel John Miller to Major George Tod, 29 July 1812.

August 1812

| Station | Recruiter |
|---|---|
| St. Clairsville | Lt. Samuel P. Booker |
| Warren | Capt. Wilson Elliott |
| Cadiz | Ensign James Milligan |
| New Lisbon | Lt. Henry Fredericks |
| Steubenville | Lt. Joseph H. Larwill |
| Zanesville | Capt. James Herron |
| Cambridge | Ensign Batteal Harrison |

On 4 August 1812 General Winchester ordered Major Tod to form two more companies.[43] Colonel Miller was ordered to march the first company from Chillicothe to Urbana, Ohio, where they would be issued their arms. Their weapons would be arriving from the military arsenal at the Newport Barracks in Kentucky.

1st Lieutenant James H. Larwill wrote to Major Tod from Canton on 9 August 1812 stating that he was sending Sergeant Jeremiah Mead with four recruits to Zanesville. Although Lieutenant Larwill was an artillery officer he had recruited a man for the infantry. He also stated that another recruit had died at his station. Lieutenant Larwill was then ordered to Steubenville, Ohio, to establish a new recruiting station.

Colonel Miller issued another order to Major Tod on 27 August 1812. The major was to send Captain Elliott and 40 to 50 recruits to Chillicothe.[44] Then on 11 September 1812 Colonel Miller issued an order to Major Tod asking him to send more recruits since Captain Elliott's company was not fully formed.[45] On 25 September 1812 Colonel Miller ordered Captain Elliott and Ensign Harrison back to recruiting service to continue recruiting men[46]. Lieutenant Larwill and his troops were ordered to Piqua, Ohio, to join Captain Daniel Cushing's company of the 2nd Regiment of U.S. Artillery.

By August 1813 a recruiting station was established at Cleveland, Ohio, and Lieutenant George Atchinson was assigned to this station as the recruiter. Major Tod was ordered to report to Colonel Miller at Fort Meigs, Ohio, during the spring of 1813. Captain Harris Hickman was temporarily placed in charge of the 2nd Recruiting District. Captain Hickman remained the acting senior recruiter until Major Tod returned from the front during the following year.

Colonel Miller was short of senior officers and he needed Major Tod at Fort Meigs. Major Tod would participate in the first siege of Fort Meigs, the invasion of Upper Canada, and he would accompany the majority of the Army of the Northwest when the army was transferred to New York after the Battle of the Thames. While in Upper Canada, Major Tod was the first American commander of the captured British fort, Fort Malden.

On 10 February 1814 new rules for recruiting in the army went into effect.[47] On this date all soldiers in the army under the age of 18 were released from duty and discharged from the army (except musicians). Recruiters could now only recruit men between 18 and 40 years of age. Previously, teenagers between the ages of 14 and 17 could join the army with the permission of their fathers, guardians or masters.

Those men re-enlisting could only re-enlist in the corps in which they originally belonged. They could not transfer from the infantry to the artillery, and so forth.

---

[43] *Western Reserve*, Tract Number 17, November 1873, (Cleveland, Ohio: 1877), page 1, Letter from Brigadier General James Winchester to Major George Tod, 4 August 1812.

[44] *Ibid.*, page 3, Letter from Lieutenant Colonel John Miller to Major George Tod, 27August 1812.

[45] *Ibid.*, page 4, Letter from Colonel John Miller to Major George Tod, 11 September 1812.

[46] *Ibid,* page 4, Letter from Colonel John Miller to Major George Tod, 25 September 1812.

[47] "New Rules for Recruiting the Army", Adjutant and Inspector General's Office, Washington, DC, 10 February 1814, *George Tod Collection*, Western Reserve Historical Society Archives Library, Cleveland, Ohio, manuscript section, call number MS-3202.

Under the Act of 3 March 1814 the 19th Infantry was consolidated with the 17th Infantry to create the new 17th Regiment of U.S. Infantry. Once the merger took effect on 12 May 1814 Major Tod was ordered back to Ohio where he once again took up the duties as the senior recruiter but this time for the new 17th Infantry. During this same time he was promoted to lieutenant colonel in the new regiment. It was his responsibility to merge the existing recruiting service of the 17th and 19th Infantries into a single service.

The recruiting returns of the 17th Infantry superintended by Captain Harris H. Hickman in the State of Ohio for the month of August 1814 is listing below.[48]

### August 1814

| Station | Recruiter |
|---------|-----------|
| Lexington, KY | Capt. Caleb Holder |
| Richmond, KY | Lt. Phillip Price |
| Maysville, KY | Lt. Jacob Anderson |
| Cynthiana, KY | Ensign Thomas Griffith |
| Louisville, KY | Ensign Rice Stewart |
| Somerset, KY | Lt. John Hamilton |
| Springfield, KY | Lt. John Taylor |
| Coshocton, OH | Lt. Jonathan Reed |
| Cincinnati, OH | Lt. George Stall |
| Cambridge, OH | Ensign William Shang |
| Steubenville, OH | Ensign William Featherston |
| Zanesville, OH | Lt. James Campbell |
| Warren, OH | Capt. James Herron |

A discharge roster showing the recruiting dates between March 1813 and May 1814 shows 41 men who were discharged from the 17th Infantry on 31 October 1814.[49] When these men were recruited, the Kentucky locations were recruiting stations for the 17th Infantry while the Ohio locations were the recruiting stations for the 19th Infantry. At the time of the discharge all of the 41 men were members of the 17th Infantry. The reporting stations and officers were:

### October 1814

| Station | Recruiter |
|---------|-----------|
| Louisville, KY | Lt. Gabriel Floyd |
| Portsmouth, OH | Ensign Hugh May |
| Detroit, Territory of MI | Lt. George Atchinson |
| Logan County, KY | Ensign John Morgan |
| Franklinton, OH | Lt. John Cochran |
| | Lt. Steven Lee |
| Frankfort, KY | Lt. Chesten Scott |
| Cincinnati, OH | Lt. Philip Price |
| Chillicothe, OH | Capt. Asabael Nearing |

After September 1814 the war in the northern part of the United States became a stalemate. The British would shift its forces to the southern states and then met its defeat at the Battle of New Orleans in January 1815. The war would end on 17 February 1815 when the British and the United States exchanged ratified treaties.

---

[48] "Recruiting returns of the 17th Regiment of Infantry", August 1814, *George Tod Collection*, Western Reserve Historical Society Archives Library, Cleveland, Ohio, call number MS-3202, container 2, folder 4.

[49] Discharge roster dated 31 October 1814 for the 17th infantry", *George Tod Collection*, Western Reserve Historical Society Archives Library, Cleveland, Ohio, manuscript section, call number MS-3202, container 2, folder 6.

Two general orders from the Adjutant and Inspector General's Office in Washington, DC, deactivated the army's recruiting service in March 1815.

The first order, dated 4 March 1815, ordered all of the officers in the recruiting service to report to their respective regiments and corps.[50] Those enlisted personnel whose enlistments expired at the end of the war were dismissed from the service while the remaining men were ordered to join their regiments and corps. Any recruit who had not joined his regiment or corps was dismissed from the army without receiving his land bounty.

The second order, dated 8 March 1815, ordered that all clothing, arms, and equipment distributed to the recruiting service would be collected and transported to one of the seven depots setup for the recruiting service.[51] Ohio's supplies would be sent to the army's Newport Barracks in Kentucky.

---

[50] Adjutant and Inspector General's Office, Washington, DC, 4 March 1815, *War of 1812 Collection*, Military Orders 1809-1815, Western Reserve Historical Society Archives Library, Cleveland, Ohio, manuscript section, call number MS-660, container 1, folder 9.

[51] Ibid, 8 March 1815, *War of 1812 Collection*, Military Orders 1809-1815, Western Reserve Historical Society Archives Library, Cleveland, Ohio, manuscript section, call number MS-660, container 1, folder 9.

# United States Air Force and the War of 1812

There is nothing wrong with this title! This is not the beginning of a spoof or the start of a bad science fiction novel. What I am going to do is to try and trace the lineage and heritage of the United State Air Force back to the War of 1812. It may seem a little far fetch but it may be fun to try and see if it can be done.

All of the branches of the United States military take pride in their units' lineage, history, heraldry and heritage. Each branch maintains an agency or department, which controls the lineage and heraldry of each of its units whether it is a land-based division, a sea-based cruiser or an air-based squadron.

Air Force Instruction 84-105,[52] paragraph 2.1.1, states, "The lineages of permanent organizations are continuous. Neither inactivation nor disbandment terminates their lineage or heraldry." If a unit is disbanded and rendered inactive the lineage will continue while the unit is idle. When a unit is reactivated, the lineage of that unit continues. The problem with lineages is that when a unit is merged into another organization then that organization inherits the lineage and history of the incorporated unit. If a new unit is later activated using the same designation as the merged unit, the new unit does not inherit that former unit's lineage.

The history and heraldry of a unit goes along with a unit's lineage. Histories are compiled quarterly and consolidated on a yearly basis. Heraldry centers on a unit's patch or emblem. Key elements in a unit's patch may have been taken from the unit's past. For example: on the 7th Infantry Regiment's emblem is a field gun signifying the Battle of Cerro Gordo during the Mexican War, where the 7th Infantry participated in the decisive attack by an assault on Telegraph Hill, a strongly fortified point.[53] When a unit merges into another unit, the new unit may combine the elements from both of the units creating a new emblem.

Heritage is more of a concept than an actual lineage. All modern U.S. Army Ranger battalions can only trace their lineage back to 1942 but they all share a common heritage that started before the Revolutionary War.[54] During most of our wars the army organized ranger companies and then disbanded them after each conflict. Although the modern ranger battalions cannot trace their lineage back to the War of 1812, the modern rangers do share the heritage of all of the ranger units that have been activated by both the U.S. Army and the Continental Army.

Up until the early 1900's Congress and the army had little desire to maintain a lineage for each of the regiments. Regiments were created as needed during wartime and then they were disbanded or merged into other regiments at the end of each hostility. Later, new regiments were created using the number designation of the old regiments.

In the history of the army there were four infantry regiments designed as either the 24th Regiment of Infantry or the 24th Infantry Regiment.[55] The first 24th Infantry was authorized in 1799 for a potential land war with France, which never materialized and the unit was not raised. The second 24th Infantry was raised for the War of 1812 and at the war's end it was merged into the 3rd Regiment of Infantry. The third 24th Infantry was authorized after the end of the Civil War when the 2nd battalion of the 15th Infantry was re-designated as the 24th Infantry. This regiment then merged into the 11th Infantry. The fourth and current

[52] Air Force Instruction 84-105, Organizational Lineage, Honors, and Heraldry, 1 March 1998, http://www.usafpatches.com/pubs/afi84-105.pdf

[53] Infantry Regiment of the United States Army, 7th Infantry Regiment, http://www.personal.psu.edu/users/j/r/jrr17/infantry/7inf.htm

[54] A Pictorial Road to a Ranger Regiment, http://www.geocities.com/Pentagon/Quarters/7632/RangerHistory.htm

[55] Heitman, Francis B., Historical Register and Dictionary of the United States Army From Its Organization, September 29, 1789, to March 2, 1903, (Genealogical Publishing Company, Baltimore, Maryland: 1994), volume I, pp. 122-124, Twenty-fourth Regiment.

24th Infantry was authorized in 1869 when the 38th and 41st Infantries merged to form this new 24th Infantry. The four different 24th Infantries do not share the same lineage or history.

All of the army's regiments merged into eight infantry regiments, a regiment of rifle, a regiment of light artillery and a corps of artillery at the close of the War of 1812. The original 1st Infantry regiment, which was formed in 1789, merged into the new 3rd Infantry, which is now the oldest infantry regiment in the army[56]. The original 2nd Infantry, formed in 1791, merged into the new 1st Infantry and this regiment is the second oldest infantry regiment in the army.[57] This all may sound a little confusing but it is necessary in order to accomplish the task at hand, which is, does the Air Force have a lineage going back to the War of 1812?

The official history of the United States Air Force traces its origins back to 1 August 1907 when the U.S. Army Signal Corps established an Aeronautical Division in order to take "charge of all matters pertaining to military ballooning, air machines, an all kindred subjects."[58] The signal corps would not receive its first airplane for another two years. On 18 July 1914 the Aeronautical Division became the Aviation Section, and then on 24 May 1918 the section became the Air Service of the U.S. Army, separate from the signal corps.

On 2 July 1926 the Air Service became the Air Corps and then on 9 March 1942 the Air Corps became the Army Air Force. This was an autonomous organization separate from the army but still under the control of the Secretary of War. The United States Air Force (USAF) was created from the Army Air Force on 26 July 1947. The USAF became a separate service with its own secretary, the Secretary of the Air Force.

Of all the numbered Air Forces, wings, groups and squadrons activated by the Air Force, the 1st Reconnaissance Squadron has the oldest lineage.[59] The army unofficially organized the 1st Aero Squadron (Provisional) at Texas City, Texas on 5 March 1913. As part of General Pershing's Punitive Expedition into Mexico, scheduled for 1914, the unit became the 1st Aero Squadron in December 1913. This 1st Aero Squadron would eventually become today's 1st Reconnaissance Squadron.

Thus, the earliest lineage established for any unit in the Air Force starts in 1913, 101 years after the start of War of 1812. But don't despair; since the Air Force was created from the Signal Corps it shares the same lineage, heritage, and history of the Signal Corps.

The first mission of the Signal Corps' Aeronautical Division was to gather military intelligence behind enemy lines. This had been one of the missions of the corps when it was first formed during the Civil War.[60] On 21 June 1860 Congress authorized the addition to the staff of the army of one signal officer with the rank and pay of a major of cavalry. The signal corps was organized under the act of 3 March 1863.[61] So now the Air Force can trace its heritage back to the authorization of a signal officer in 1860.

The airplane would become just one more tool for the signal corps to use in support of its mission. The signal corps, in turn, was an offshoot of the cavalry since the first signal officer held a position created from the cavalry. Actually, this was a dragoon position since cavalry regiments would not be designated until 1861.

---

[56] Ibid, pp. 85-87, Third Regiment.

[57] Ibid, pp. 81-83, First Regiment.

[58] The Air Force Historical Research Agency, Organizational History Branch, The Birth of the United States Air Force, http://www.maxwell.af.mil/au/afhra/wwwroot/rso/birth.html

[59] Global Security.ORG, Military, 1st Reconnaissance Squadron (1st RS), http://www.globalsecurity.org/military/agency/usaf/1rs.htm

[60] John Patrick Finnegan, Military Intelligence, (Center of Military History, United States Army, Washington, D. C.: 1998), Chapter 1, The Beginnings.

[61] Heitman, Historical Register, volume I, page 44, Signal Corps.

The Regiment of Dragoons was organized under the act of 2 March 1833.[62] The regiment became the 1st Regiment of Dragoons after the organization of an additional regiment under the act of 23 May 1836. The designation of the regiment was changed to 1st Cavalry Regiment under the act of 3 August 1861.

From this, the Air Force can now trace its lineage back to the creation of the first regiment of dragoons in 1833. With the heritage of both the signal corps and the cavalry under its belt, it is easy to see why the early Air Force operated in a very similar matter to that of the cavalry.

"Early in the twentieth century, military doctrine treated air operations as an extension of the cavalry -- in effect a sky cavalry."[63] For example, a January 1912 report to the French Chamber of Deputies argued that "the aeroplane should not replace the cavalry, even in reconnaissance work; its action should be auxiliary to that of [the cavalry] and complete it." Echoing this sentiment in 1913, Brigadier General George P. Scriven, Chief Signal Officer of the U.S. Army, testified before Congress that "the aeroplane is an adjunct to the cavalry." Even as late as 1920 a much celebrated U.S. Army Air Service regulation seemed to reflect cavalry connections: "Pilots will not wear spurs while flying!" The earliest tactical organization developed around the airplane was the squadron, a term which was borrowed from the cavalry.

To go beyond 1833 to the War of 1812 now becomes a little more difficult since there were no cavalry-type units in the army between 1815 and 1833. The Regiment of Dragoons was technically a different organization than the Regiment of Light Dragoons that was disbanded in 1815.

The light dragoons (light cavalry units) had traditionally been the main intelligence gathering organization for the army during both the Revolutionary War and the War of 1812. Light dragoons were designed for reconnaissance, screening missions and transporting official messages. They were also used to disrupt the enemies supply lines and they could skirmish with the enemy on horseback.

The traditional cavalry regiments were trained to fight on horseback against the enemy's cavalry and to charge the enemy's battle lines disrupting the infantry. Cavalry regiments were not created until the Civil War. Dragoons were mounted infantries that used horses for transportation and then they dismounted in order to fight. Another difference between the cavalry, the light dragoons and the dragoons was the type of weapons that each of these elements used.

The three dragoon regiments that were authorized beginning in 1833 were actually hybrid regiments, which performed the traditional duties of a light dragoon regiment, a dragoon regiment and a true cavalry regiment. These three types of cavalry units are still classified as 'cavalry'. The Air Force, in a manner of speaking, did inherit the heritage of the cavalry and the various types of cavalry units.

The Air Force's lineage starts with the formation of the first dragoon regiment in 1833 and continues through the creation of the signal corps in 1863. The Air Force cannot claim the lineage of the light dragoon regiments of the War of 1812 since these units merged into the Corps of Artillery on 17 May 1815.[64] From the signal corps is born the aviation units of the early 20th century, and follows with the evolution of air power until the United States Air Force was created in 1947.

Although the Air Force's lineage starts in 1833 it has inherited the heritage of both the cavalry (in all of its forms) and the signal corps. Thus the Air force has inherited the history and heritage of the cavalry from the War of 1812.

---

[62] Ibid, pp. 65-66, First Regiment of Cavalry.

[63] The Air Force Historical Research Agency, Organizational History Branch, A Guide to United State Air Force Lineage and Honors,
http://www.maxwell.af.mil/au/afhra/wwwroot/rso/guide_usaf_lineage_honors.html

[64] Heitman, Historical Register, volume I, pp. 79-80, Regiment of Light Dragoons.

## War of 1812 POW resources at the Ohio Genealogical Society Library

Over the past few years the Society of the War of 1812 in the State of Ohio has donated microfilms containing the records of American Prisoners of War held by the British during the War of 1812. This is a very unique set of records that are usually not found in American libraries.

The seven microfilms contain copies of the British Admiralty's General Entry Books of American Prisoners of War which are held in the Public Record Office in London, Great Britain.

The records are from the following prisoner of war camps: Chatham, Dartmoor, Plymouth, Barbados, Bermuda, Halifax, Cape of Good Hope, Jamaica, Newfoundland, Odiham and Quebec. They contain information on 28,125 American soldiers, sailors, marines, privateers and civilians who were held by the British during the war.

The General Entry Book records are composed of lines for the recording of names of those incarcerated. The record of each prisoner is found on two facing pages.

The columns across the top of the left side:
Current Number (prisoner's number)
By What Ship, or how taken
Time When – Day, Month, Year
Place Where
Of What Ship or Corps
Whether Man of War, Privateer or Merchant Vessel
Prisoners' Names

The columns across the top of the right side:
Quality (prisoner's military rank)
Time when received into Custody – Day, Month, Year
From what Ship or whence received
Exchanged, Discharged, Died, or Escaped
Time When – Day, Month, Year
Whither, and by what order, or Number of Re-entry

Some of the ledgers contain additional information on each prisoner. These ledgers list the place of birth, age, color of hair and eyes, and marks or wounds. If a prisoner had been transferred from one prison to another, then an additional entry for clothing and bedding would be noted on the ledger if that prisoner had brought the items with him.

The records of the Chatham prison in England has the listing of 3,955 POWs interned between October 1812 and October 1814. The records of the Dartmoor prison in England has the listing of 6,553 POWS interned between April 1813 and March 1815.

The records of the Barbados prison has the listing of 1,453 POWs interned between August 1812 and March 1815. The records of the Bermuda prison has the listing of 2,435 POWs interned between July 1812 and January 1815.

The records of the Halifax prison has the listing of 8,148 POWs interned between June 1812 and January 1815. The records of the Cape of Good Hope prison has the listing of 296 POWs interned between November 1812 and April 1815.

The records of the Jamaica prison has the listing of 1,553 POWs interned between July 1812 and March 1815. The records of the Newfoundland prison has the listing of 364 POWs interned between July 1812 and November 1812.

The records of New Providence prison has the listing of 836 POWs interned between July 1812 and March 1815. The records of the Quebec prison has the listing of 1,990 POWs interned between June 1813 and December 1814.

There were four additional American POW microfilms that have not been purchased yet by the Society of the War of 1812 in the State of Ohio.

# Index

Buckingham, Jared 122
Bundy, William 122
Burge, Thomas 77
Burgoyne, John 115
Burr, Aaron 18
Burrow, Jarrel 58, 60
Burrow, Philip 58, 60
Burt, John 122
Bush, Henry C. 123
Bush, Stephen 81
Bushnell, Andrew 123
Bussey, Herman 123
Butler, John 123
Butson, William 77
Byrnes, Timothy 123
Caffry, John R. 70
Cagle, Simon 58, 60
Cain, Robert 123
Caldwell, John 48
Caldwell, John W. 123
Callender, Henry B. 81
Camm, James M. 123
Campbell, James 123, 148
Campbell, William P. 123
Camper, Jonathan 70
Campfield, Jesse 81
Cannon, William S. 123
Cantlen, Martha J. 26
Carena, Charles 123
Carle, Peter 123
Carlisle, John 123
Carney, David L. 20
Carpenter, Dorman 123
Carpenter, Isaac 47, 48
Carpenter, Nathaniel 77
Carpmail, William 77
Carr, John 48
Carrel, Philip 22
Carroll, Anthony Wayne 23
Carroll, Catherine 23
Carroll, Henry 23
Carroll, Jane 23
Carroll, John 23
Carroll, Joseph 23
Carroll, Margaret 23
Carroll, Mary 23
Carroll, Philip 23
Carter, Landon 38
Carter, Mary 8
Cartledge, Samuel 77
Casebeer, David 30, 31

Caskey, John M. 123
Cass, Charles Lee 19
Cass, Lewis 114
Catharell, Joseph 123
Cavan, Timothy 123
Cavender, Sarah 8
Chamberlain, Hannah 8
Chamberlane, John 77
Champlin, Stephen 85
Chandler, John 68
Chapman, James 49
Chapman, Jonathan 70
Chase, Nathaniel 81
Chauncey, Isaac 73
Clachan, Alexander 123
Clark, Adam G. 123
Clark, John 76, 123
Clark, Miss 8
Clark, Samuel 123
Clark, William 17
Clarke, Charles 58, 60
Clinton, Thomas 81
Coates, Daniel 70
Cobb, Cyrus 81
Cobb, Michael 123
Cochran, James J. L. M. 70
Cochran, John 148
Coffee, George 123
Cokely, Cornelius 77
Cole, Samuel D. 123
Cole, Thomas 32
Coleman, William 123
Collison, William 77
Conklin, Augustus H. M. 85
Conley, John R. 59, 60
Connotchy, Philip 31
Conyers, John 103
Cooke, Levin 70
Copeland, Weeks 123
Copps, Darius 49
Copps, Josiah 123
Cork, William 77
Correll, Mary Ann 34
Cotton, Joshua T. 35, 36
Cowen, Andrew 59, 60
Cox, John 77
Cox, William 123
Craig, George 70
Cral, Jeremiah 123
Crane, Joseph L. 70
Crany, Theodore 123

Crapin, Samuel 81
Crawford, Thomas 49
Creach, Thomas B. 49
Cremens, Moses 123
Crenlieu, William 56
Cressey, Moses 123
Crooks, Richard 91, 92
Crosby, Jeremiah T. 103
Crubbs, Philip 123
Crump, ----- 8
Culins, George 123
Cummings, John L. 16
Cunningham, Joseph 123
Currant, James 77
Currey, Rebecca 24
Curry, George 24
Curry, Isaac 24
Curry, Robert 123
Cushing, Daniel 6, 147
Cutright, Catherine 40
Cutright, John 40
Dailey, John 49
Dailey, Peter 59, 60
Dains, Andrew 123
Daley, John 70
Damon, Zachariah 81
Daniels, Benjamin 123
Daniels, George W. 103
Davis, Ebenezer 84
Davis, Justice 103
Dawdle, Richard 77
Dawes, James 70
Dawson, William 123
Day, Cornelius 70
Dearborn, Henry 69, 79
Deaton, Thomas 78
Deets, Sarah 24
Delahay, Henry 70
Dell, John 49
Denio, Frederick 123
Deolin, Patrick 78
Deppisen, John C. 70
Derby, Abraham 81
Derr, Mathias 123
Devault, John 123
Devon, Enos 123
Dibble, Asa 81
Dickey, George 31
Dickson, Robert 123
Dill, John 56
Divin, Hugh 24

Gragg, Enus   82
Graham, Jesse   51, 56
Graham, John   124
Grammer, Peterson   58, 61
Grammer, Pleasant   58, 61
Graves, Edward   77
Gray, Amey   8
Gray, Henrietta   40
Grayson, John   71
Greaves, John   56
Green, Elisha B.   124
Greenlee, Mrs.   8
Greeny, Mrs.   8
Griffin, Mrs.   8
Griffin, William   73, 74
Griffith, Thomas   148
Grinton, Martin   21
Grinton, Philip   21
Grinton, Robert   21
Griswold, Henry W.   43
Guiles, Eleanor   7, 9
Guiles, John   7, 9
Guiles, Joseph   7
Hall, Samuel   124
Hall, Silas   51
Hamilton, John   51, 148
Hamilton, Samuel C.   124
Hammond, Cynthia   9
Hand, Elijah   56
Haning, Mrs. Aaron   6, 9
Hanna, John   68, 71
Hanscom, Eleazer   51
Hanson, Samuel   51
Hardy, Reuben   51
Harker, Mary   9
Harper, William   124
Harrington, Jason   51
Harriott, John   58, 61
Harris, William   58, 61
Harrison, Batteal   145, 147
Harrison, Elizabeth   9
Harrison, William Henry   14,
   28, 29, 57, 89, 91, 114
Harriss, Ann   9
Hartgrave, Francis   58, 61
Hartshorn, David   82
Harvey, Samuel   51, 56
Haskell, William   51
Hatch, Ward   82
Hatford, Samuel   82
Havens, Joel   124

Hawkins, Richard   78
Hayes, George   71
Hayes, Nicholas   71
Hayley, Edward   71
Haymaker, John   58, 61
Hays, Joseph   58, 61
Hazeltine, Thomas   69, 71
Hazelton, Ann   9
Hearn, Daniel   82
Hedges, Solomon   58, 61
Helmick, Mary   32
Henderson, John   52, 58, 61
Henn, Charles   125
Henry, Catharine   9
Henry, Polly   9
Henson, Samuel   56
Herndon, John   82
Herrald, Cader   58, 61
Herron, James   144, 145, 146,
   147, 148
Hetick, Mrs.   9
Hickman, Harris M.   22, 147
Hickman, Rosanna   9
Higgins, Daniel   56
Higgins, James   52
Higgins, Robert   58, 61
Higgins, William   125
Hill, Jeremiah   52
Hill, John   125
Hill, Levi   56
Hills, Joseph   56
Hilton, James   71
Hines, William   58, 62
Hinkson, Thomas   92
Hitchcock, Peter   37
Hoagland, Enoch   103
Hodges, William   58, 62
Hoefer, Henry (alias Andrew)
   125
Hogan, Daniel   125
Hogg, Elijah   52
Holbrook Jr., John   58, 62
Holbrook Sr., John   58, 62
Holden, William G.   103
Holder, Caleb   148
Hollister, Jesse W.   125
Hollowell, Benjamin G.   84
Hollowell, Betsey   9
Hollyfield, William   59, 62
Holmes, Andrew Hunter   89
Holmes, Elizabeth   9

Holmes, James   125
Honser, Henry   125
Hood, Margaret   9
Hook, Joseph   67
Hooper, William   58, 62
Hotchkiss, Henry   82
Hough, Enoch   32
Houtzell, Jacob   82
Houzer, Susanna   34
Howard, Benjamin   14, 58
Howard, John   71
Howard, William   28
Howland, Elisha W.   82
Hoyt, Fitch   58, 62
Hughes, William   125
Hull, Abraham Fuller   115
Hull, Isaac   115
Hull, William   91, 115, 145, 146
Humphrey, Mrs.   9
Humphries, Thomas   59, 62
Hunt, Peggy I.   9
Huntington, Samuel   27
Hurst, Charles B.   56
Hussy, Edward   56
Hutchins, James   103
Hutchins, Loomis   56
Hutchinson, Joseph   52
Hutchinson, Seth   52
Hutchinson, Sewell   52
Hutchinson, William   52
Ingalls, Abigail   7, 9
Ingalls, Amos   7
Ingalls, Stephen   7, 9
Irvine, Baptist   68, 69, 71
Isbell, Ransom   125
Jack, James   125
Jackson, Andrew   14, 17, 89
Jackson, George W.   145
Jackson, John   92, 125
Jackson, William   125
Jarvis, William H.   71
Jefferson, Thomas   17
Jennison, John S.   125
Jesup, Thomas Sidney   115
Johnson, Abraham   125
Johnson, Arnold   125
Johnson, Eve   9
Johnson, Hetty   9
Johnson, Joseph   30
Johnston, Harvey   125
Jones, Becky   9

McIntire, Joseph C.  126
McKinzie, William  58, 63
McLaughlin, Robert  126
McNabb, John  126
McNulty, James  126
Meeker, Thomas J.  126
Meese, Henry  32
Meigs Jr., Return Jonathan  27, 37, 89
Meins, John  126
Melcher, Isaac  71
Mellus, Daniel C.  53
Melona, Cornelius  82
Mengro, Joseph  19
Merchant, Richard  71
Merian, James F.  53
Merian, Samuel  53
Metz, Charles  126
Meyer, Henry  126
Mick, Adam  58, 63
Mickle, Adolph  126
Middleton, Abner  77
Milburne, Andrew  126
Miles, Jesse A.  126
Miller, Charles  91, 92, 126
Miller, George  126
Miller, Jacob  126
Miller, James  115, 147
Miller, John  146
Miller, Joseph  24
Miller, Peter W.  103
Miller, Philip  53
Milligan, James  146, 147
Milligan, John  145
Millikin, James  82
Millikin, Sterling F.  82
Mills, Nathaniel  126
Mincher, William  53
Minshall, John  126
Mitchell, James  58, 63
Mitchell, John B.  126
Momeny, George  126
Monk, Charles  56
Monroe, James  89
Monshene, William  56
Montgomery, Shadrack  126
Moody, Thomas  56
Moore, Obadiah  53
Moore, Stephen H.  67, 69, 71
Morgan, Catherine  26
Morgan, John  148

Morgan, Sally  10
Morris, Mary  10
Morris, Pressley  41
Morrison, George W.  126
Morrison, Moses  53
Morrison, Robert  145
Morse, Amos  126
Mortimer, Benjamin  29, 30
Morton, Thomas  126
Moss, Abigail  10
Mullett, Isaac  54
Mullinax, Elijah  19
Mullinax, James  19
Mullinax, John  20
Murphy, Sarah  10
Murphy, William  126
Myers, John  71
Myers, Samuel  126
Nash, John  126
Nearing, Asabeal  148
Newburgh, John V.  71
Nicholas, Jacob  54
Nichols, David  126
Nichols, Josiah  82
Nichols, Ransom  58, 63
Nichols, William  85
Nickerson, Betsey  10
Noon, Darby  67
Noonan, Michael  126
Norton, Andres  73, 74
Norton, Margaret (or Mary)  10
Noyes, Jacob  82
Nutting, Cyrus  54
Nutting, Simri  54
Nye, John  126
O'Kelly - see Elizabeth Kelly
Oldfield, Sophia  23
O'Neil, Gregory  78
Orkins, James  54
Ormsby, Nancy  10
Orr, Alexander  58, 63
Orwick, Amos  26
Orwick, Henry Clay  23
Orwick, James H.  26
Orwick, Joseph Palmer  23
Orwick, Margaret  24
Orwick, Mary  23
Orwick, Matilda  26
Orwick, Samuel  22
Orwick, Sarah Jane  24
Orwick, Silas  26

Orwick, William T.  24
Owens, Thomas  126
Packett, John  85
Page, Charles  83
Page, David  54
Page, George A.  126
Page, Joseph  56
Paleifer, Joseph  54
Palmer, Joseph  23
Palmer, Lavina  24
Palmer, Sarah  23
Palmer, Thomas  73, 74
Pamken, James  77
Parker, Ezekiel  58, 63
Parker, James  83
Parker, John  54
Parks, George W.  126
Parks, Laban  58, 63
Parnell, Jacob  58, 63
Parris, Alexander  79, 83
Parrish, Abraham L.  103
Partridge, Flavel  85
Paskiel, Ezekiel D.  54
Patterson, Alexander  126
Patterson, John  71
Paul, David  54
Payett, Mary  11
Payson, Edward  83
Pea, Mrs. John  11
Peak, Elijah  103
Pearson, Abel  58, 63
Peate, W. F.  59, 63
Peck, David  32
Peckham, Polly  56
Peeler, Pleasant  63
Penman, John  71
Peregoy, William  71
Perkins, Simon  37, 39, 91, 143
Perry, Anna  7, 11
Perry, Calvin  7
Perry, Mrs.  11
Perry, Oliver Hazard  73, 85, 87, 89
Perry, Parmelia  7, 11
Peters, John  73
Peters, John  74
Peters, William  71
Phelps, William  63
Philips, Joseph  57, 58, 63
Philips, Thomas  64
Phillippa, Cecilia  11

Smith, Giles  78
Smith, Isaac  103
Smith, James  58, 65
Smith, John  78, 127
Smith, Joseph  57
Smith, Joseph D.  58, 65
Smith, Richard  127
Smith, Samuel B.  58, 65
Smith, Sara  26
Smith, Thomas  28
Smith, William N.  83
Smyth, Alexander  68
Sneed, John  58, 65
Snelling, Josiah  115
Snyder, William  127
Son, Jacob  103
Soule, James  128
Sparks, Elizabeth  12
Sparks, Richard  17
Speakes, Edward L.  72
Spears, William G.  128
Spohn, John  128
Spoonover, James  128
Sprague, Frederick A.  83
Springer - see Montgomery
  Shadrack
Spunagle, Samuel  103
St. Clair, Arthur  17
Stagg, Daniel  128
Stall, George  148
Stevens, Elhaman  128
Steward, Marvin  128
Stewart, Aaron  128
Stewart, Rice  148
Stewart, Thomas  58, 65
Stigall, Zachariah  58, 65
Stiles, Timothy S.  55
Still, Samuel  128
Stoddard, David  128
Stollings, Martin  128
Stone, Mary  12
Stoner, John  12
Stoner, Mary  6
Storey, John  58, 65
Storm, John  128
Storrs, Stephen  83
Stoutsberger, Andrew  72
Stratton, Willis  58, 65
Stubbs, James R.  43
Stultz, Adam  6
Stultz, Fanny  6, 12

Sturtivant, Martin  83
Subre, Elizabeth  40
Suit, Susanna  12
Sullivan, Timothy  83
Summeranter, Francis  128
Summers, John  58, 65
Swan, Rachael  12
Swinburn, Johanna P.  12
Swyers, Daniel  128
Symmes, Charles  58, 65
Symmes, John  28
Taggart, James  83
Tayler, Mrs.  12
Taylor, Jemima Anne  18
Taylor, John  148
Taylor, Michael  128
Taylor, Samuel  58, 65, 128
Taylor, William  128
Teater, John  5
Tennile, Benjamin  57, 58, 65
Terleman, John  128
Teypool, Ennis J.  128
Thomas, Edward  83
Thomas, Moses  83
Thomas, Robert  83
Thompson, Elizabeth R.  26
Thompson, Samuel  128
Thornburg, Jesse M.  128
Tiffany, Cyrus  85
Tinham, Thomas  78
Tod, George  143, 144, 145,
  147, 148
Todd, Rebecca  12
Tolly, James  58, 65
Tope, Abraham  32, 34
Tope, Elizabeth  32, 34
Tope, George  33
Tope, Henry  32
Tope, John  32, 33
Tope, Levi  32, 34
Tope, Susanna  34
Townsend, Jonathan  83
Towson, Nathan  67
Tracey, William  72
Trask, -----  12
Trimble, Allen  27
Truly, Samuel  128
Truman, Charles Y.  128
Tucker, Mary  12
Tucker, Robert  58, 65
Tufts, Samuel  72

Tupper, Edward W.  91, 92
Turner, Daniel  85
Turner, Edward  128
Turner, Ezekiel A.  128
Underwood, James  72
Van Burgen, William D.  72
Van Dyne, Isabella  12
Van Horne, Isaac  31
Vanatta, Isaac  128
Vance, John  128
Vance, Joseph  27
Varnum, Jacob B.  55
Varnum, Samuel M.  56
Voorhise, Luke  128
Wadkins, Darius  128
Wadleigh, Thomas  56
Walcutt, William  87
Waldron, Richard  83
Walker, Jane  12
Walker, Mrs.  12
Walker, Seth  56
Wallace, Nicholas  128
Wallender, Herman  128
Walton, Mathew  103
Wardle, John  78
Warner, Thomas  67, 72
Warnick, Robert  103
Warnock, Thomas  128
Waterhouse, Jacob  84
Waters, Polly  6, 12
Watkins, Thomas  78
Watkins, William  72
Watson, Joseph  128
Wayne, Anthony  40
Webb, William  72
Weber, Henry  128
Weedon, John H.  128
Weenesdoefer, Lawrence  128
Weir, Ann  7, 12
Weir, David  7
Weir, Lucinda  7, 12
Welch, Moses  72
Welch, Robert  128
Wentling, Ann  12
Westlake, James  78
Wheelock, Ezra  84
Whicher, Samuel E.  84
Whicher, William  84
Whipple, Lucius  128
Whistler, James  115
Whistler, John  115

Heritage Books by the Society of the War of 1812
in the State of Ohio:

Transcribed by Harrison Scott Baker

*American Prisoners of War Held at Bermuda,*
*Cape of Good Hope and Jamaica During the War of 1812*

*American Prisoners of War Held at Barbados,*
*Newfoundland and New Providence During the War of 1812*

*American Prisoners of War Held at Halifax*
*During the War of 1812, Volume I and II*

Transcribed by Eric Eugene Johnson

*American Prisoners of War Held at Dartmoor During the War of 1812*

*American Prisoners of War Held in Montreal*
*and Quebec During the War of 1812*

*American Prisoners of War Held at Plymouth*
*During the War of 1812*

*American Prisoners of War Held at Quebec*
*During the War of 1812, 8 June 1813–11 December 1814*

*American Prisoners of War Paroled at Dartmouth,*
*Halifax, Jamaica and Odiham During the War of 1812*

*American Sea Fencibles in the War of 1812:*
*United States Sea Fencibles, State Sea Fencibles*

*Black Regulars in the War of 1812*

*Black Regulars and Militiamen in the War of 1812*

*Forgotten Americans Who Served in the War of 1812*

*Ohio and the War of 1812: A Collection of Lists, Musters and Essays*

*Ohio's Regulars in the War of 1812*

Heritage Books by the Society of the War of 1812
in the State of Maryland:

*Maryland Regulars in the War of 1812*
Transcribed by Eric Eugene Johnson; Foreword by Christos Christou

www.ingramcontent.com/pod-product-compliance
Lightning Source LLC
Chambersburg PA
CBHW080614270326

41928CB00016B/3050